(24 HOURS)

Please Print Clearly Within the Boxes

AF573845

Month Day Year

Total Miles Driving Today

CARRIER NAME & ADDRESS

Tractor Number

Driver's ID / Code

Driver's Signature in Full

I certify these entries are true and correct

Trailer Number

Co-Driver's ID / Code

Co-Driver's Name

MID-NIGHT	1	2	3	4	5	6	7	8	9	10	11	NOON	1	2	3	4	5	6	7	8	9	10	11	TOTAL HOURS
1: OFF DUTY																								.
2: SLEEPER																								.
3: DRIVING																								.
4: ON DUTY (NOT DRIVING)																								.

MID-NIGHT 1 2 3 4 5 6 7 8 9 10 11 NOON 1 2 3 4 5 6 7 8 9 10 11

REMARKS:

1/4 = 25
1/2 = 50
3/4 = 75

MADE WITH PRIDE IN THE U.S.A.

Returned to normal work location at end of day

Pro Number

\# of Days Off Duty Includes Today

Pre-Trip Inspection Signed

Post-Trip Inspection Signed

Original File at home terminal
Duplicate Driver retains in his/her possession for eight days

DRIVER'S DAILY LOG

(24 HOURS)

Will be Scanned
Please Print Clearly Within the Boxes

Month Day Year

Total Miles Driving Today

CARRIER NAME & ADDRESS

Tractor Number

Driver's ID / Code

Driver's Signature in Full

I certify these entries are true and correct

Trailer Number

Co-Driver's ID / Code

Co-Driver's Name

MID-NIGHT	1	2	3	4	5	6	7	8	9	10	11	NOON	1	2	3	4	5	6	7	8	9	10	11	TOTAL HOURS
1: OFF DUTY																								.
2: SLEEPER																								.
3: DRIVING																								.
4: ON DUTY (NOT DRIVING)																								.

MID-NIGHT 1 2 3 4 5 6 7 8 9 10 11 NOON 1 2 3 4 5 6 7 8 9 10 11

REMARKS:

1/4 = 25
1/2 = 50
3/4 = 75

MADE WITH PRIDE IN THE U.S.A.

Returned to normal work location at end of day

Pro Number

\# of Days Off Duty Includes Today

Pre-Trip Inspection Signed

Post-Trip Inspection Signed

Original File at home terminal

As required by the Federal Motor Carrier Safety Regulations Part 396.11

Date ________ Tractor No. ________________ ☐ I detect NO defect ☐ I detect the following defects:

INDICATE DEFECT AND SHOP, CITY & STATE WHERE REPAIRED.

____

____

OPERATOR'S SIGNATURE	MECHANIC'S SIGNATURE	SIGNATURE OF OPERATOR REVIEWING REPAIRS

TRAILER VEHICLE CONDITION REPORT

TRUCK NO. ________________

Date ________ Trailer No. ________________ ☐ I detect NO defect ☐ I detect the following defects:

INDICATE DEFECT AND SHOP, CITY & STATE WHERE REPAIRED.

____

____

OPERATOR'S SIGNATURE	MECHANIC'S SIGNATURE	SIGNATURE OF OPERATOR REVIEWING REPAIRS

TRACTOR VEHICLE CONDITION REPORT

As required by the Federal Motor Carrier Safety Regulations Part 396.11

Date ________ Tractor No. ________________ ☐ I detect NO defect ☐ I detect the following defects:

INDICATE DEFECT AND SHOP, CITY & STATE WHERE REPAIRED.

____

____

OPERATOR'S SIGNATURE	MECHANIC'S SIGNATURE	SIGNATURE OF OPERATOR REVIEWING REPAIRS

TRAILER VEHICLE CONDITION REPORT

TRUCK NO. ________________

Date ________ Trailer No. ________________ ☐ I detect NO defect ☐ I detect the following defects:

INDICATE DEFECT AND SHOP, CITY & STATE WHERE REPAIRED.

____

____

OPERATOR'S SIGNATURE	MECHANIC'S SIGNATURE	SIGNATURE OF OPERATOR REVIEWING REPAIRS

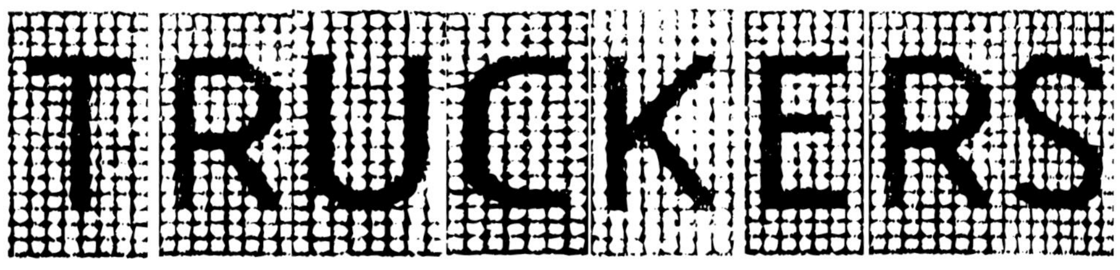

BY Mary Richardson

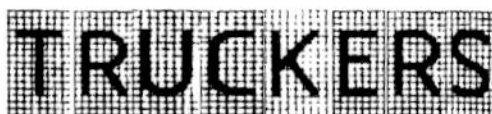

by Mary Richardson

All photographs © their respective photographers.
For a complete list, see page 127.

Design: Carolyn Frisch & Christopher D Salyers
Production Director: Christopher D Salyers
Editing: Buzz Poole

Typeset in: Nilland

Library of Congress Control Number: 2008929891

Printed and bound in China by Asia Pacific Offset

10 9 8 7 6 5 4 3 2 1 First edition

Mark Batty Publisher
36 West 37th Street, Suite 409
New York, NY 10018

www.markbattypublisher.com

ISBN: 978-0-9799666-8-2

Distributed outside North America by:

Thames & Hudson Ltd
181A High Holborn
London WC1V 7QX
United Kingdom

Tel: 00 44 20 7845 5000
Fax: 00 44 20 7845 5055

www.thameshudson.co.uk

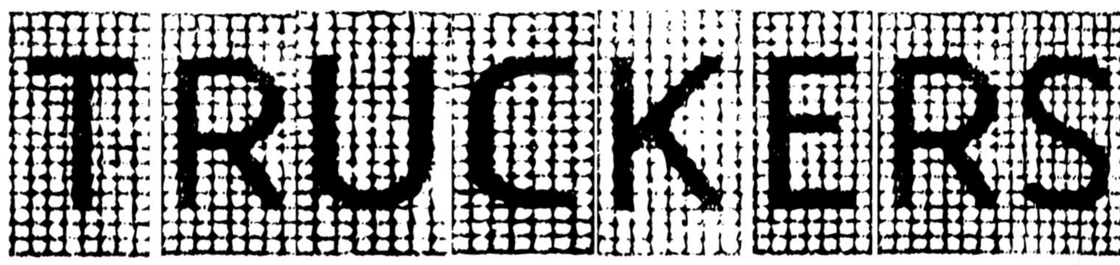

TRUCKERS

BY Mary Richardson

WITH PHOTOGRAPHY BY

PHIL ANDREWS

JENNY WILLIAMSON

MESHAKAI WOLF

Mark Batty Publisher

New York City

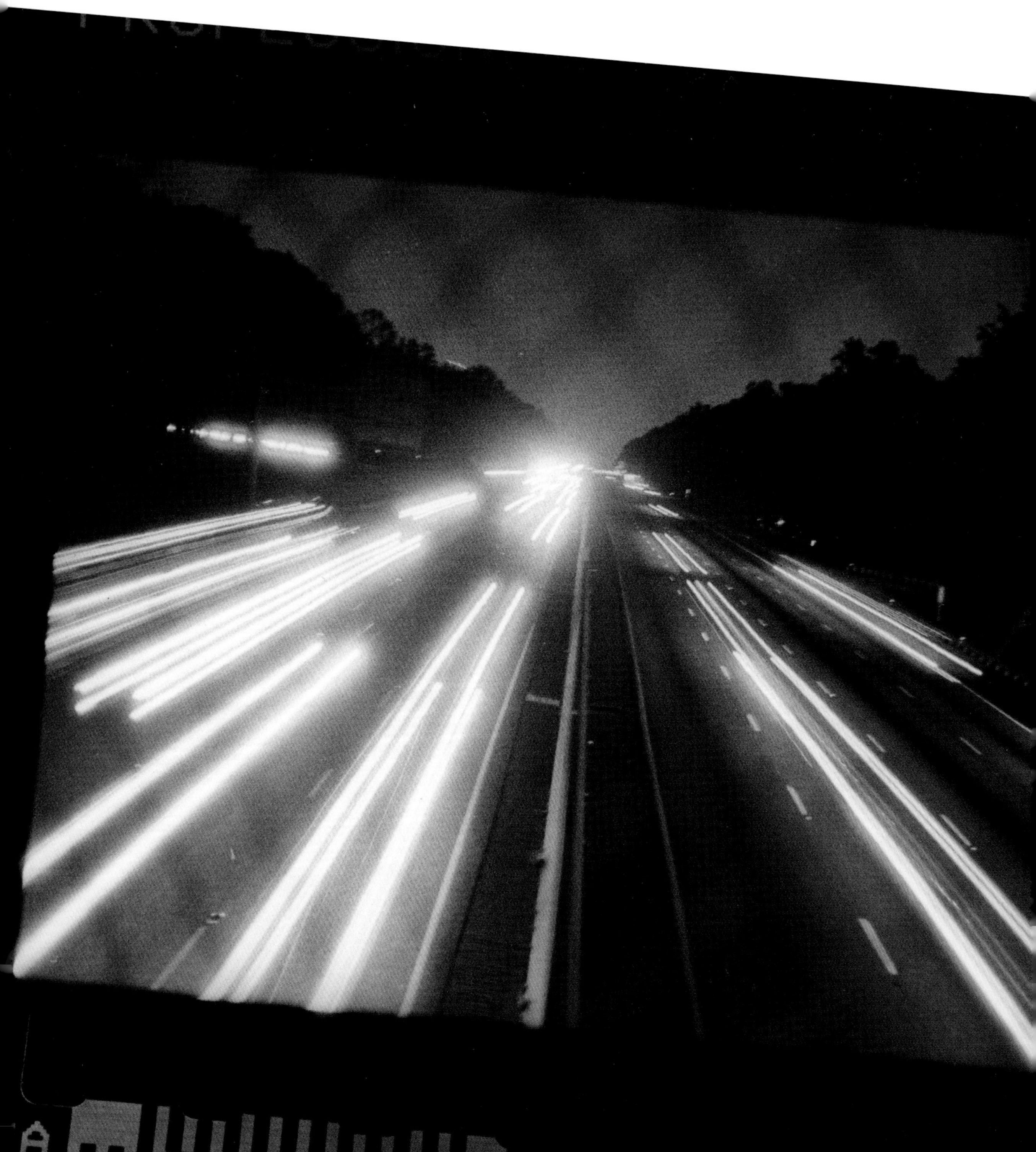

To those whom we owe our comforts

Gratitude and love to Benjamin Solomon, Andrea Sanders, Misty Harper, Jimmy Lo, Teri Jenkins, James Clay, Rebecca Heinegg, James and Suzanne Andrews, Dianna Settles, Ashley Andrews, Trucker's Connection, Mecca & Son's, Wonderroot, Donna's Brothel, Buzz Poole, Christopher D Salyers, Mark Batty and all the truckers we had the pleasure to meet.

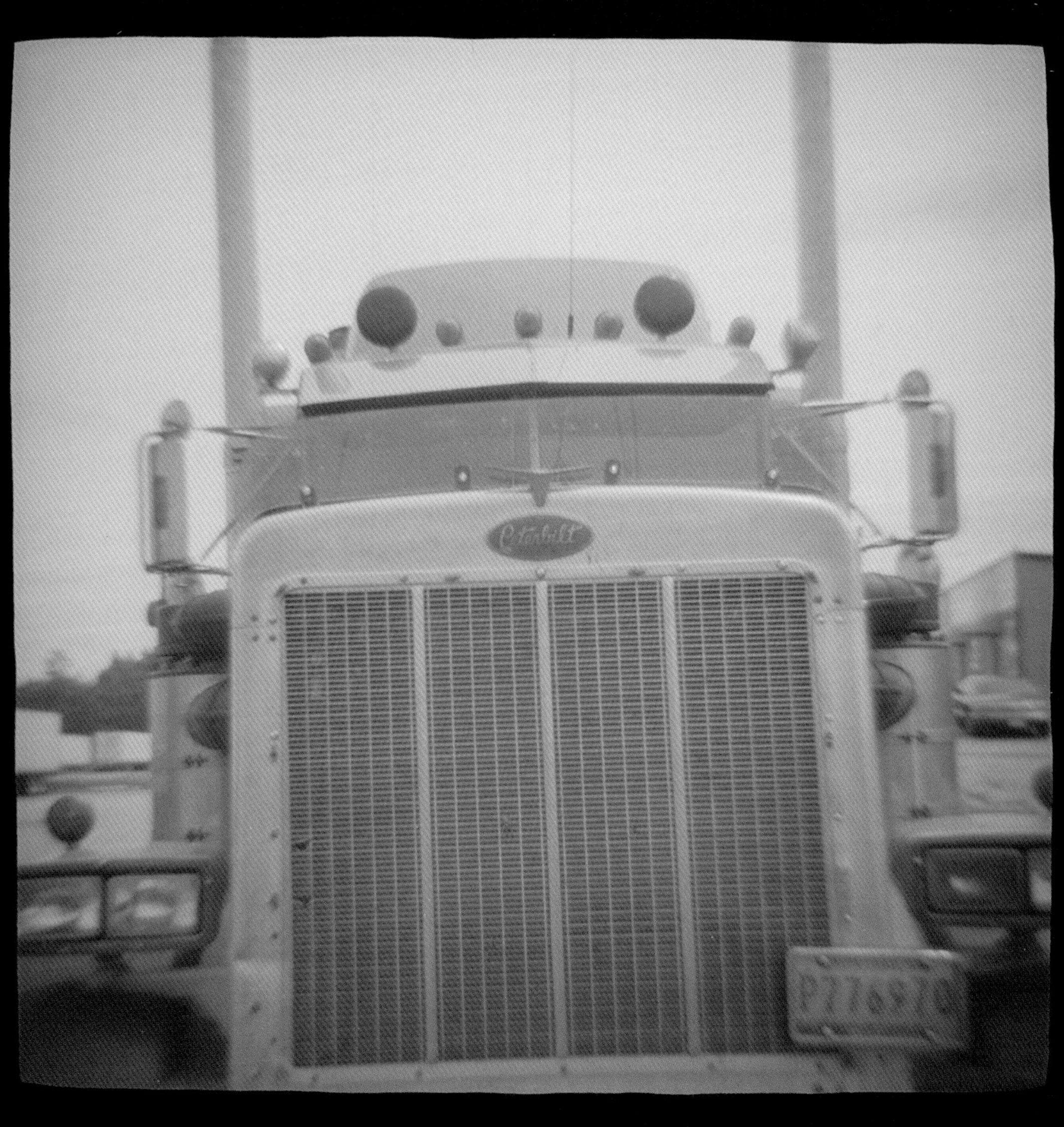

TABLE OF CONTENTS

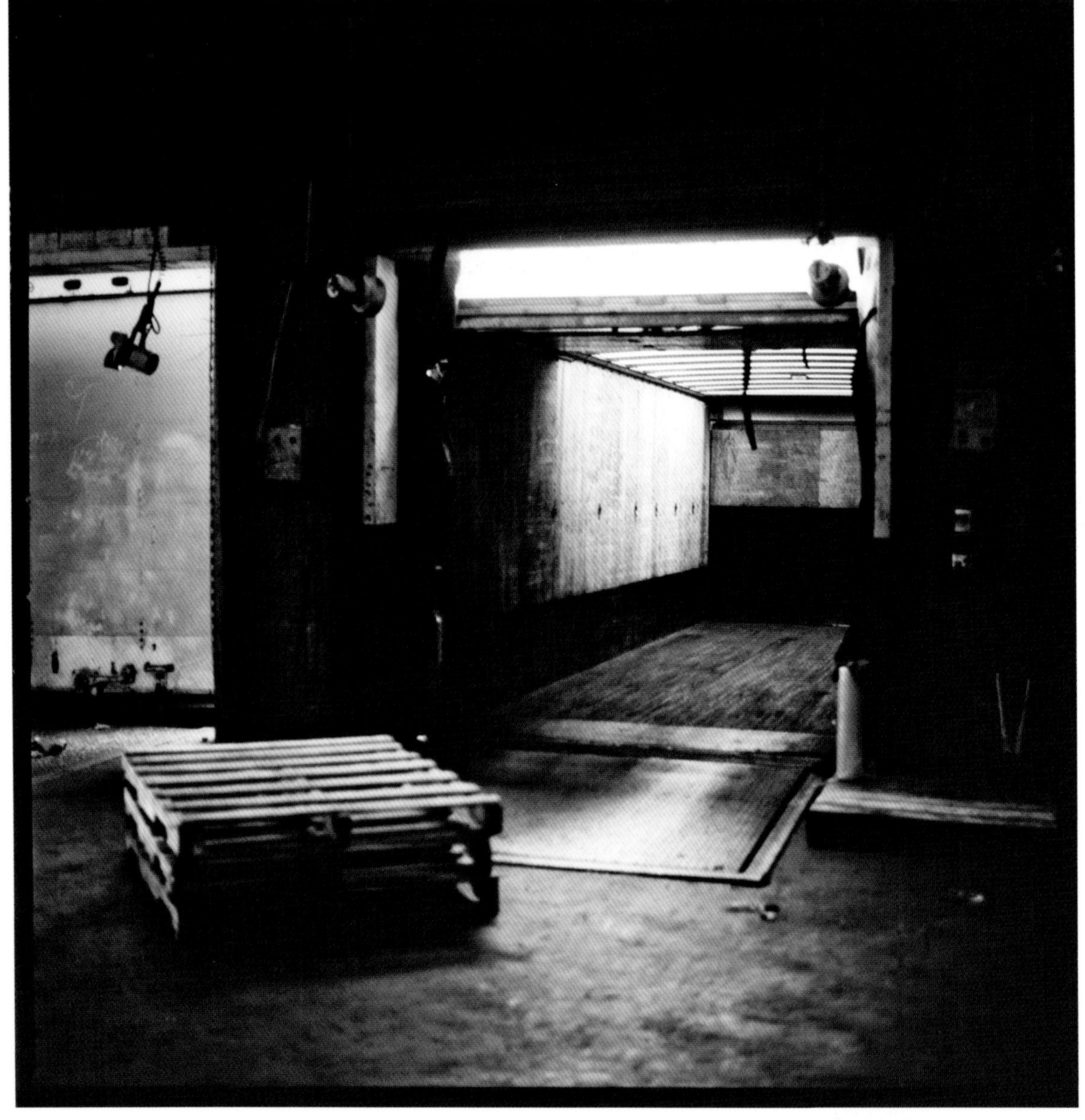

Coffee warehouse; Jersey City, New Jersey

EVOCATION

The deception is that working longer, working harder, giving up our lives for this work will pay off. It is a road that leads us in circles, leaving us unable to free ourselves from debilitating patterns. *Most of the truckers interviewed for this book are on the road two, three, four months at a time, returning home for two or three days before leaving again.* There is a time for working hard, but this clutter, traffic, homogeny and ceaseless motion seem to break less and less, the cycle hardening its establishment in the framework of our society.

Truck drivers are realized human machines of the industrial age, if there were such a thing as a human machine molded of human hands, every movement shaped by a human demand, shaped by heads spinning on bodies, spinning in a series of related tornadoes. And below their certain shyness something still radiates because they want to be heard. But there is no room in their design for their desires.

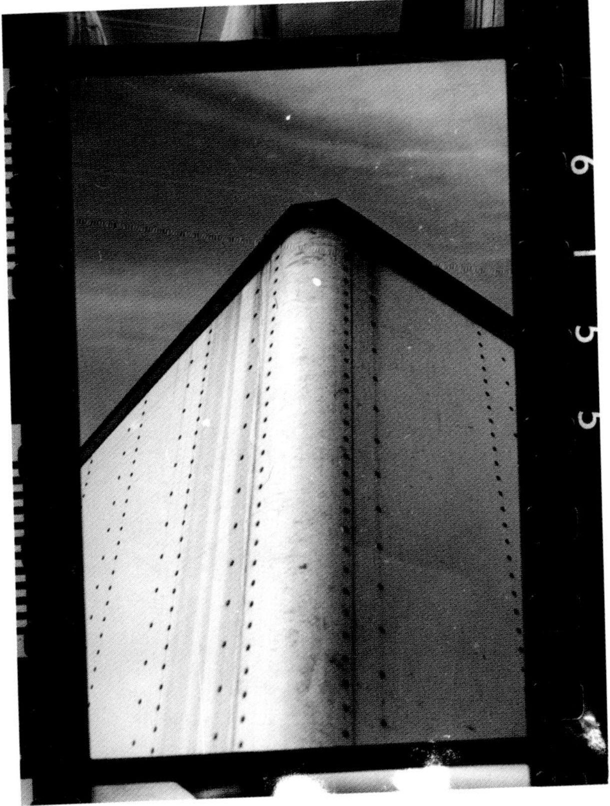

Truck trailer; I-85 Exit 53

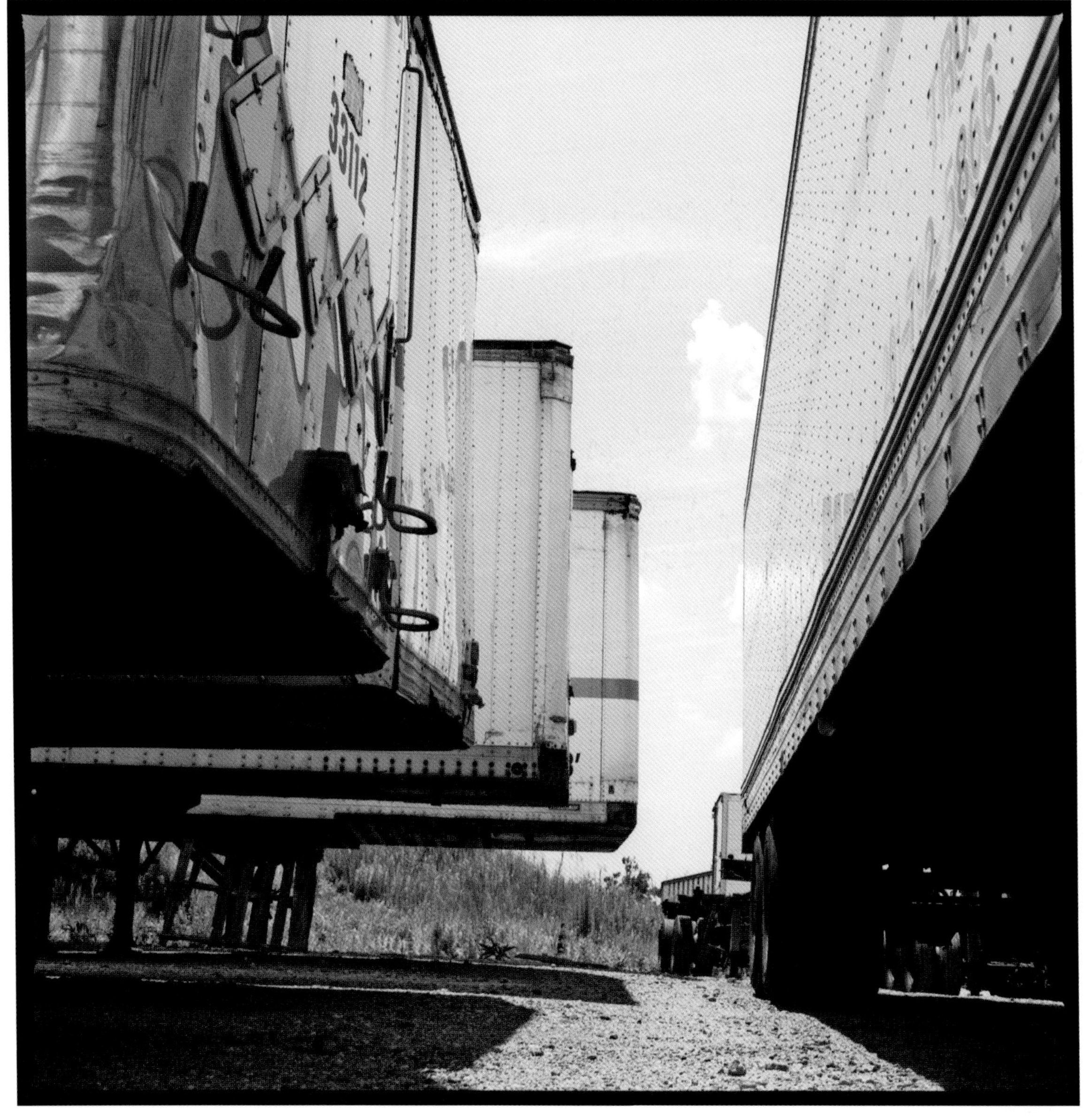

Parked trailers

A mayonnaise, on its journey

By Benjamin Michael Solomon

INGREDIENTS: SOYBEAN OIL, WATER, WHOLE EGGS AND EGG YOLKS, VINEGAR, SALT, SUGAR, LEMON JUICE, NATURAL FLAVORS

Mayonnaise begins near Urbandale, Iowa, in acres and acres of soybeans diligently plumping themselves to ripeness in the late summer sun. Harvested and pulverized, the soybeans are extracted into a crude and odorous soap that is refined, sparged with water and finally pumped into a five-thousand gallon tanker truck: pure, edible soybean oil. Christian Burns, father of four, drives the tanker 311 miles to the Blue Bonnet Mayonnaise Factory in Joliet, Illinois, where the oil rests in a vat just below room temperature, awaiting its creamy destiny.

Popped from the bodies of chickens in Gainesville, Georgia, the eggs shuffle and teeter en mass through the processing facility, perfect fragile packets of protein and cholesterol. They are cleansed of their glistening effluvium by a gentle spray and paraded before the sharp-eyed handlers, who send them skidding down an aluminum slide to be packaged in cartons of 100 eggs each. These are driven 767 miles to Joliet by Steven Birches, high school graduate, who leaves his delicate cargo in the Blue Bonnet Mayonnaise Factory's half-acre cold storage warehouse.

Old as the earth itself, the water has no beginning, but we encounter it now sucked from the plentiful aquifer beneath Joliet into the Eastmoreland Water Service Association Treatment Plant where it is fluorinated, chlorinated and released through 14 miles of sluices and pipes in total darkness to gush forth triumphantly from industrial spigots into 100-gallon stainless steel vats.

Giant mechanical mixers stir the water into the soybean oil, the whole eggs and the egg yolks, the vinegar (which came from corn near Mansfield, Ohio, shipped 345 miles to Altoona, Pennsylvania, was distilled into alcohol, fast-fermented and hauled 566 miles to Joliet by Milton Graves, Army Reservist), the salt (hauled from a mine 1,095 miles away in Hockley, Texas, by Jason and Marion Best, husband and wife team), the sugar (from sugar beets grown near Moorhead, Minnesota, and transported 655 miles by Ernst Walker, Republican) and the juice of lemons (from Grove City, Florida, via a pasteurization facility near Tampa, 457 miles by train to Atlanta and loaded onto the rig of Steven West, independent contractor who drives 726 miles to Joliet in a personal record of just under 9 hours, non-stop, peeing in a bottle, Lynyrd Skynyrd blasting all the way).

Of the "natural flavors" we know nothing.

Mixed and commingled, vigorously beaten and slowly frothed into a creamy gel, the mayonnaise is pumped through an intricate network of rubber tubing into a pasteurization tub, heated and cooled, re-pumped into pressurized vats and, portion by portion, expunged into brightly labeled plastic squeeze-bottles.

(Squeeze-bottles began deep beneath the ground, sucked from the bowels of the earth in Iran as crude oil, refined in a petrochemical plant into resin pellets of high-density polyethylene and shipped 3,962 miles overseas to Shanghai. Ho Jin, born-again Christian, hauled the pellets 126 miles to Taizhou City, China, "Kingdom of Plastic Products," where they were heated to a creamy sludge, bleached white and poured into 24-ounce squeeze-bottle molds. Guang Yum hauled them 126 miles back to Shanghai, where they shipped 5,331 miles to the US in a special consignment of 30 containers aboard a Dutch freightliner, skippered by Captain Cornelis Slordt. It docked in San Francisco, disgorged its cargo onto the North American continent and each container made its way roughly 2,500 miles across the country by routes both common and varied, in trucks driven by men both old and young, some with wives in small towns, some with girlfriends in cities, some in convoys, many alone, some in cabs with refrigerators, some with televisions, some angry, some wired, some horny, some sad, all bound for The Blue Bonnet Mayonnaise Factory in Joliet, Illinois, where the enterprise of mayonnaise production employs just under one-sixteenth of that small city's population.)

Nearly full to the brim with creamy white mayonnaise, one bottle is capped, wrapped in a protective plastic seal, trundled down the conveyor line and loaded by prehensile mechanical arms into a cardboard box numbering 45 units. Warehouse employees load the boxes onto a truck driven by Walter Harrison, divorced, who hauls it west, 2,248 miles to McKinleyville, California, where one bottle of mayonnaise sits on the grocery shelf, waiting, waiting, until Sunday, July 24, when Agnes Pritchard has a craving for potato salad.

Gone in a single afternoon down the hungry gullets of Agnes and her husband Pete, the mayonnaise has completed its journey. But the trusty squeeze-bottle, world traveler, Iranian born Chinese national, is destined to continue wandering. After sitting in a bin in Agnes and Pete's garage for a week, it is loaded (by Pete) into the trunk of their Prius with a variety of other plastics and deposited at the recycling center, where it is crushed and compressed into the middle of a ten-foot cubic bale and stored for 40 days until a 12-ton truckload has accumulated. Kenneth Brown, one-fourth Native-American, hauls the bales 303 miles to a recycling exporter in Sacramento, where they endure both sun and rain until plastic brokers in Hong Kong purchase the bales and ship them 5,989 miles back to China.

Jianguo Chang, chain-smoker, drives the bales 174 miles from Hong Kong to a village near Guangzhou in the Guangdong Province. Here workers break open the bales, scrape away the now-faded label and throw the flattened squeeze-bottle of Blue Bonnet Mayonnaise into the grinder, which devours it and spits out the shards. These are heated in a vat, melted to liquid and reconstituted as 100% synthetic-fiber filler for a sleeping bag that four months hence, by routes circuitous but inevitable, finds its way 6,000 miles back to the US, again, and ships 2,194 miles across the country to a discount sporting good store in Grenada, Mississippi, to be purchased one winter afternoon by Christian Burns, father of four, former trucker.

The actual distance traveled by the components in this bottle of mayonnaise is 33,879 miles. The circumference of the earth at the equator is 24,901 miles.

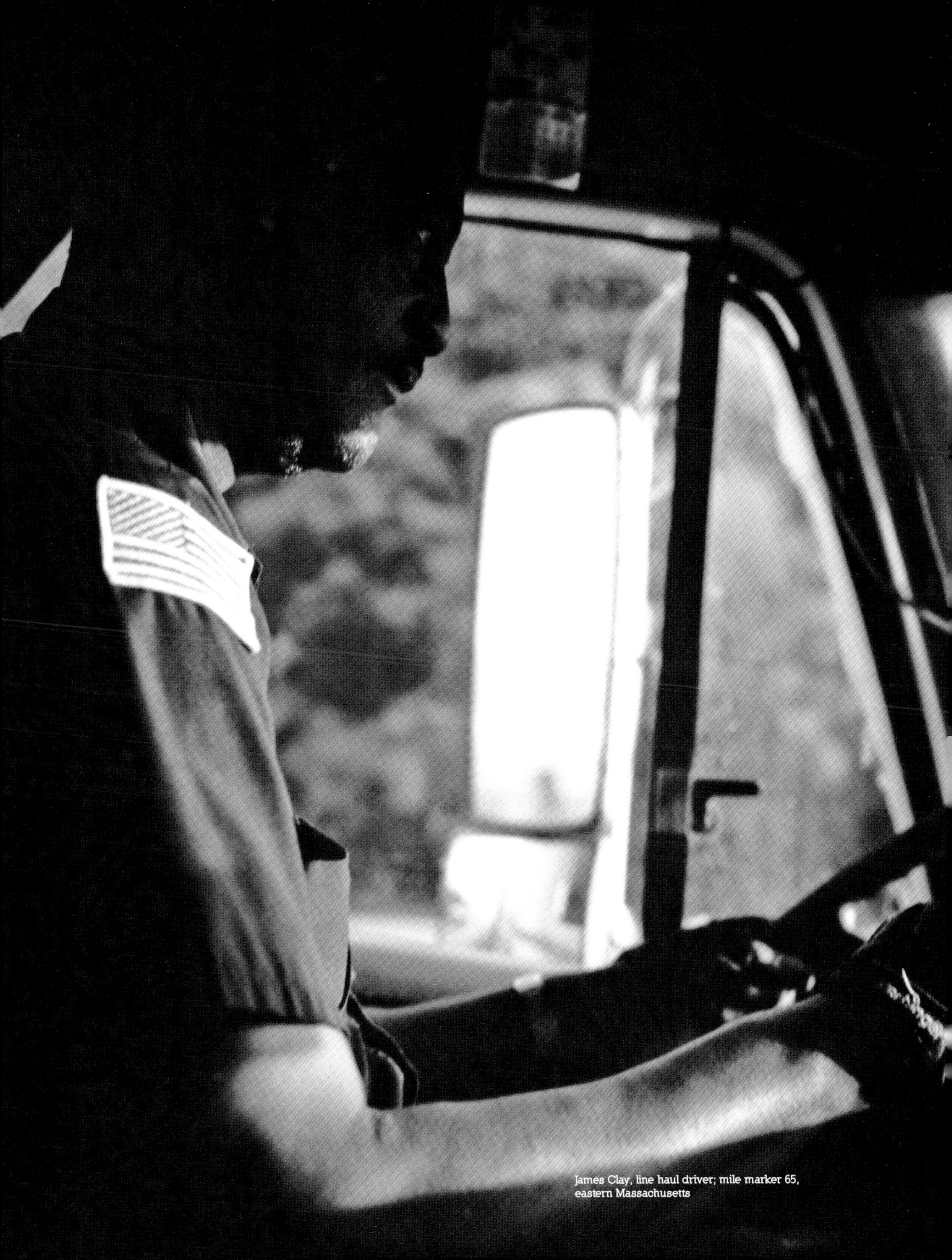

James Clay, line haul driver; mile marker 65,
eastern Massachusetts

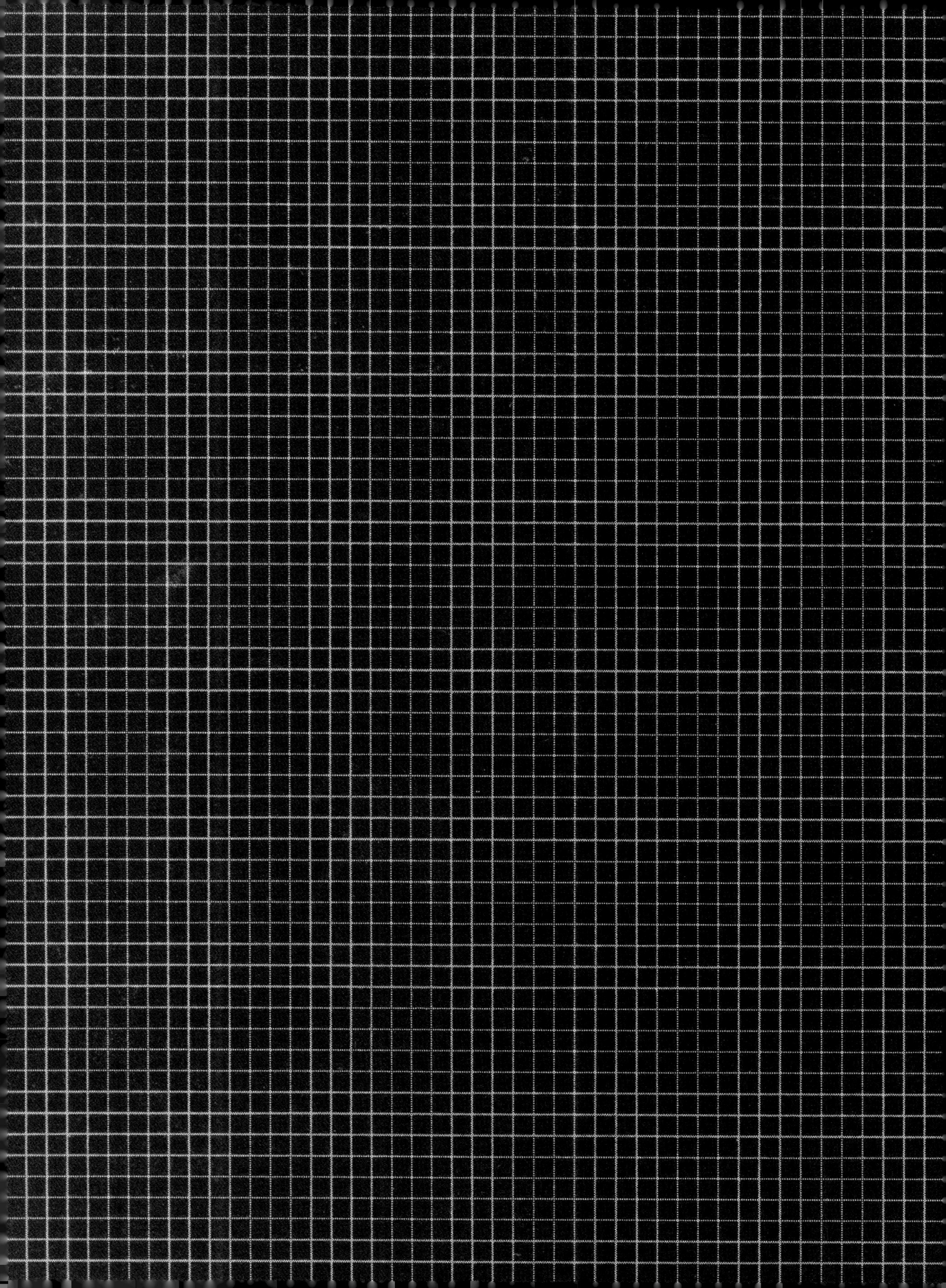

PART ONE
Understandings, isolation, sleep

An Understanding

In his hand are two brands of cigarettes; the one between his index and middle fingers has a brown filter, the one between his ring and middle fingers, a white filter. One is smoked down, the other about halfway. His hand hangs down over his knee and the smoke swirls up over his knuckles to his wrist. Duct tape holds the seal along the edge of the window. He sits sideways with his feet propped on the step beneath the cab, staring into the distance. His beard is patched together, black, white and grey. There is no scenic landscape beyond signs and buildings placed sporadically: wheel shines, truck wash, repairs, Sunday service, showers. He faces it, relaxed and unfocused, as he has for 36 years. Tension is absent in this moment; it is the breath before a journey indefinite.

Truck Stop, I

From certain angles, the Pilot Travel Center looks like a red plastic Lego block floating in a puddle of grey. Beyond the rear door, the metallic roar of engines trudging up the hill encompasses all other sound; in its consistency it goes unnoticed ... *paper, television sets, water bottles, cardboard boxes, auto parts, air conditioners, soup, plastic bags, Proctor & Gamble products, bricks, Kellogg's cereals, cars, diapers, plywood, steel, sheetrock, rubber, tomatoes, red meat, toilet paper, air ducts, hairspray, computers, mustard, potatoes, frozen dinners, motors, lawnmowers, circus equipment, crude oil, pipe, livestock, paint, propane, canned vegetables* ... the weather is wet, bone chilling. Nick Juno is an oddity existing quietly apart yet united with a mass of sameness. He is smiling, shaven, not yet grizzled but wearing a rumpled shirt. He exclaims that he has seen every kind of bird there is while out on the road. On the dashboard lies his copy of *366 Readings from Buddhism*. He started driving three months ago and has not been home yet. For 15 years he was a car salesman. He is fond of the open road, nature and independence: "It's great. No boss is looking over your shoulder, it's a chance to travel and see the country, and it's an opportunity to earn a living based on how hard you work."

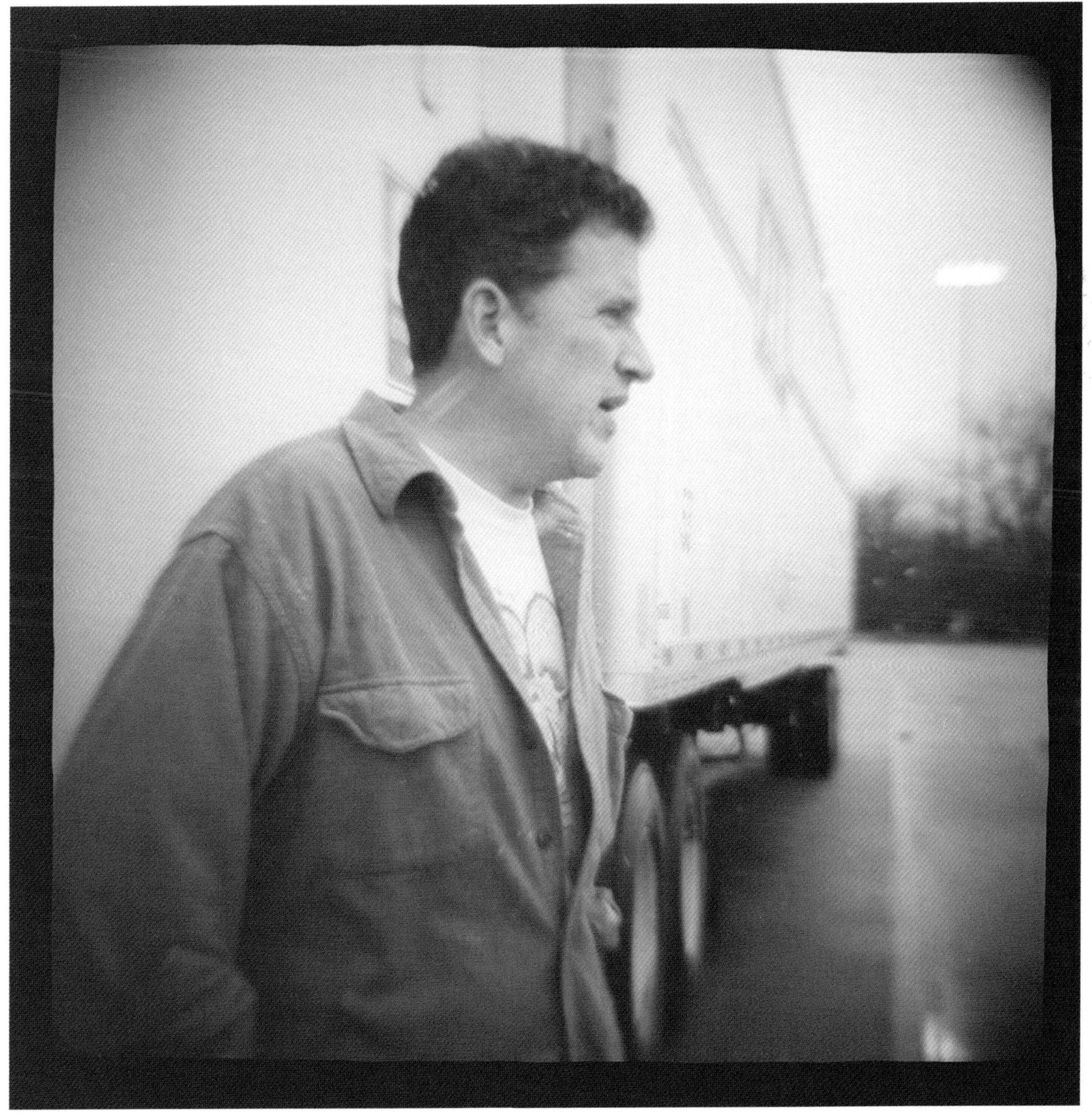

Nick Juno; I-20, Exit 114
b.1963
Divorced, 2 daughters

"I Seen a Picture of Him In 1992."

A young woman drives to the warehouse to meet him. He gets there early and they are both happy that his wait is long. Time passes. Finally, papers in one hand and his other arm around her shoulder, they walk out of the office. Except for everything before and after this hour, the day is romantic. Time passes. He gets into the truck and she follows him in her car until he has to turn off. She is crying. Time passes. Babies are conceived and born in the blinking gaps between. Time passes. A ten-year-old daughter lives in Guadalajara with relatives. Time passes. An eighteen-year-old son he's never met agrees to meet him somewhere when there's time.

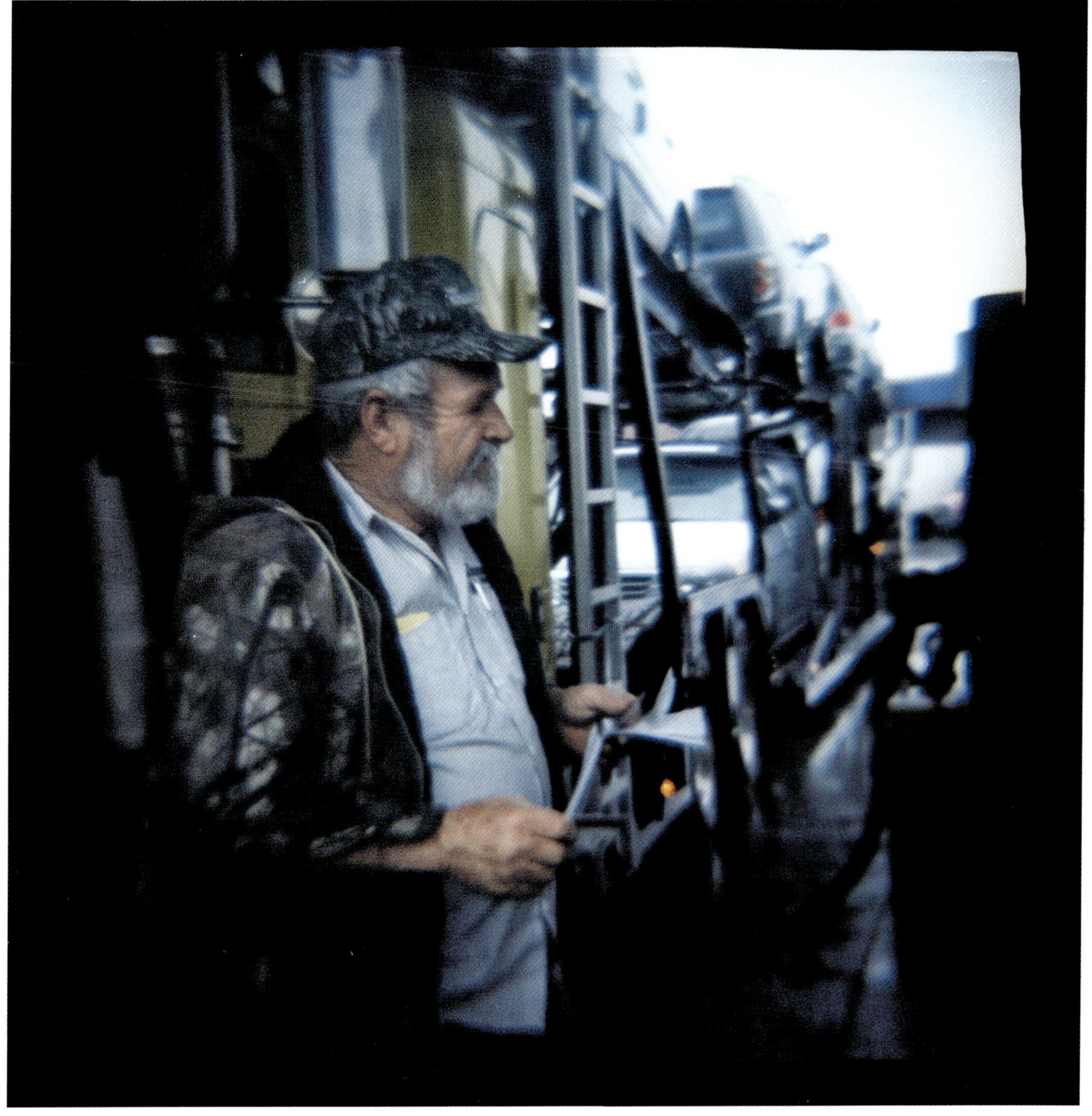

Joe Nelson, hauling cars; I-20, Exit 114

Analogue

It is estimated that there are more than 1.8 million truckers in the US.

It is estimated that the industry needs 20,000 more.

The shortage is estimated to reach 111,000 by 2014.

It is estimated that 30 year's work is equivalent to 4,500 freight containers.

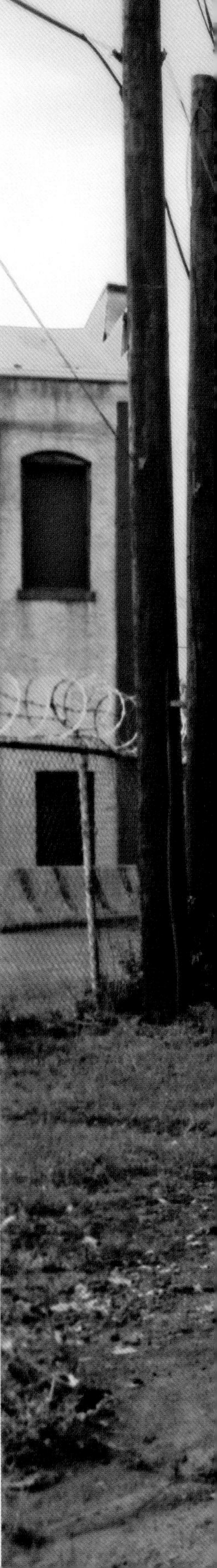

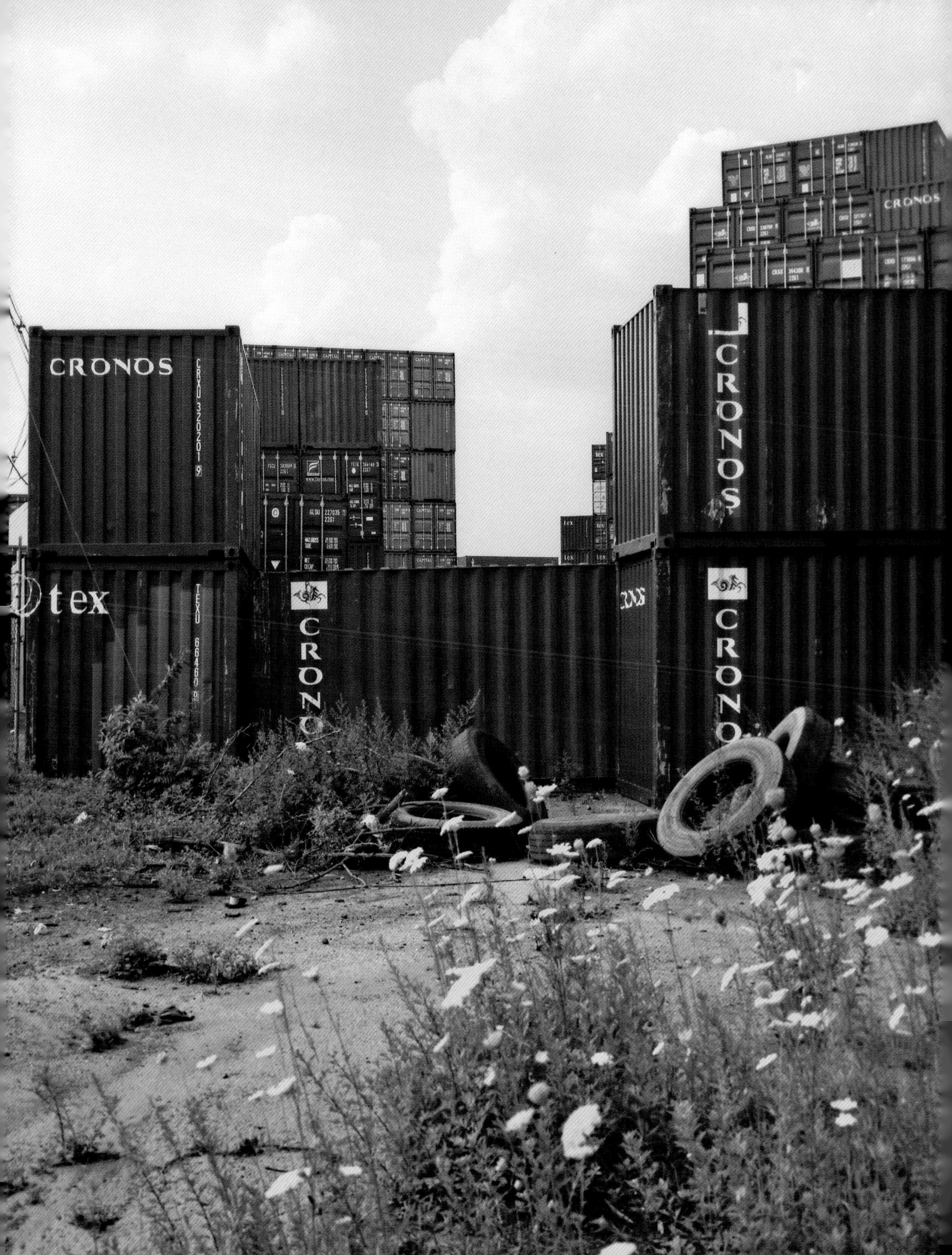
CRONOS
CRXU 320201 9
tex
TEXU 66460 9
CRONOS
CRONOS
CRONOS
CRONOS
CRONOS

Litmus Test

The American Trucking Association's truck tonnage index is measured every three months. Economists use it to predict recessions and consumer spending; it is the litmus test of the economy. Seventy percent of manufactured goods purchased in the US travel by truck, and the trucking industry regularly absorbs 60 to 80 percent of the money spent on all modes of freight transport. In 2006, $645.6 billion was spent on 10.7-billion tons of truck freight.

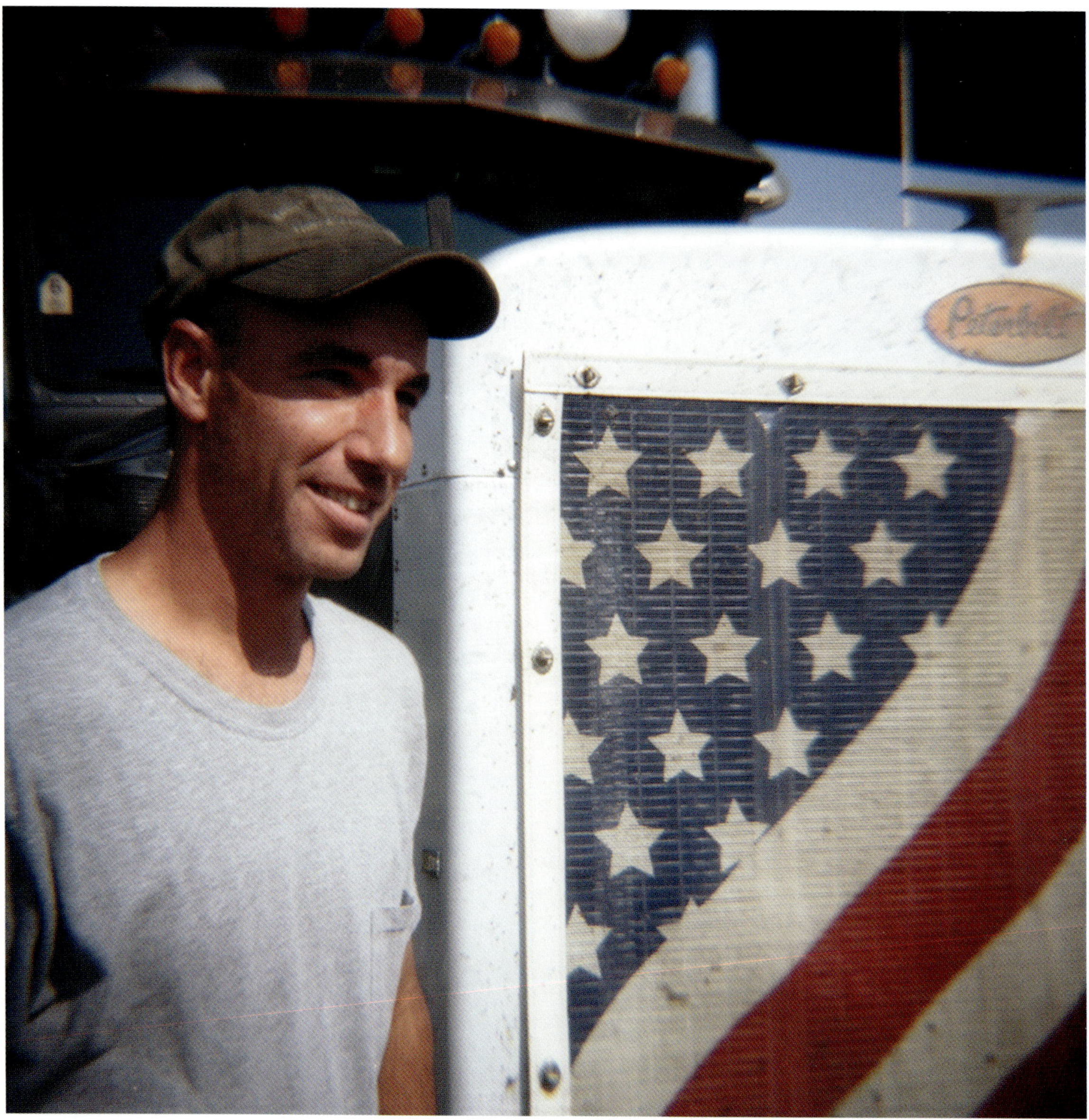

Ike Nollmeyer
Years driving: 13
Odometer: 700k

Current load: structural steel for a new business center

In a Cab, I

Butch Rogers's feet and head touch the walls when he lies on the bed in the bottom bunk; a folding scooter is stored on the top bunk for transportation once the truck is parked. His blanket and sheet lie crumpled in the corner and stuffed animals are strapped into the passenger seat, all gifts from his wife.

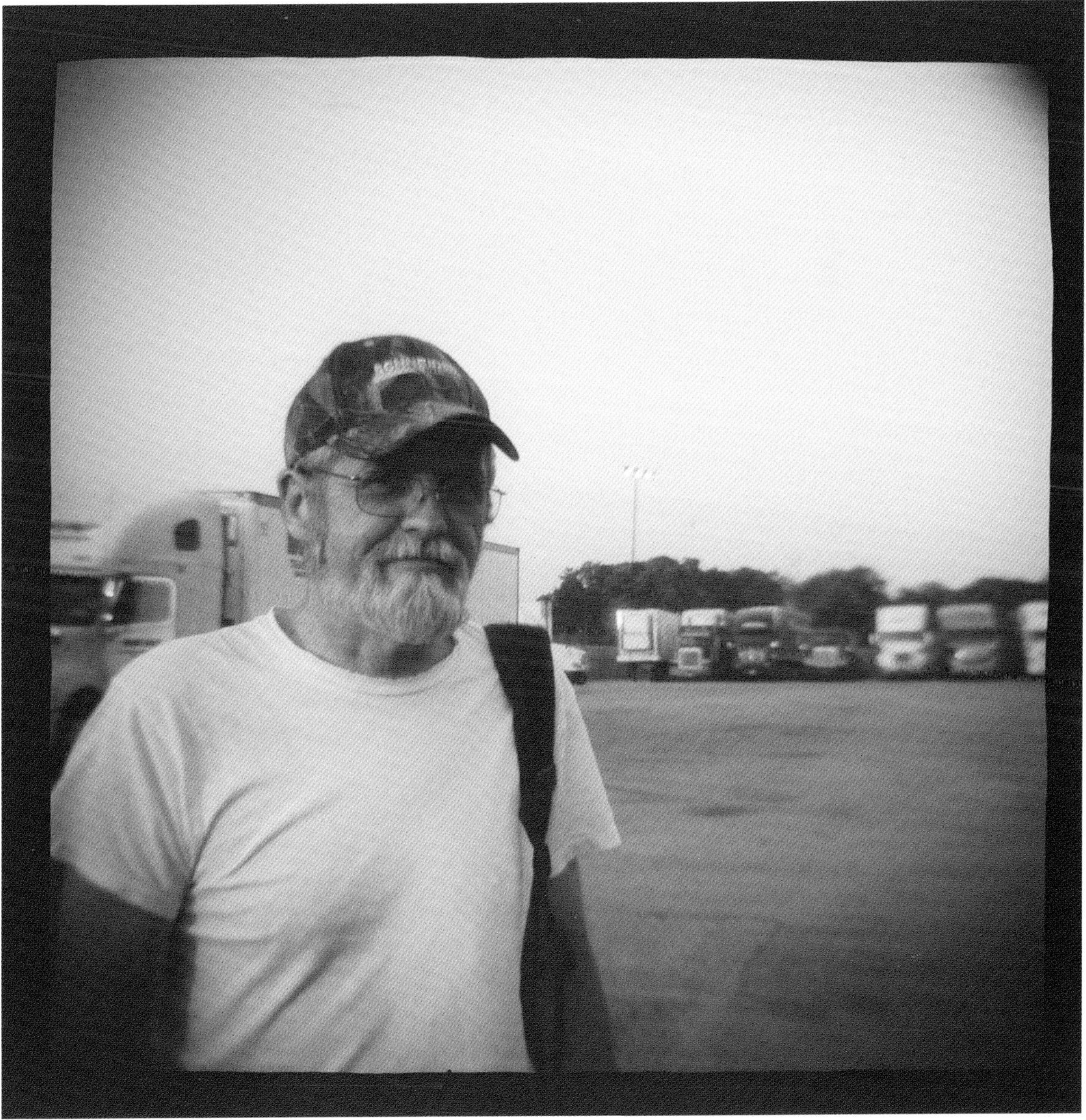

Trucker with backpack; I-85, Exit 129

A Hard Road

Jeff Lipfort is parked for the night, having a beer. "I seen a woman burn in her car because I couldn't get the safety belt undone. Myself and a cop tried to get her out. I quit driving for a year after that." Last week, he was ticketed because someone stole his pigtails, the lines that run from the cab to power trailer lights. He has lost many things over the years; he lost his wife and home in 2002. "Truck driving caused me to get a divorce. It was a good two-and-a-half years. All I can say is that she couldn't stand me being gone that long." Besides homes and families, truckers are at a constant risk of losing a lot more. If they have an accident with a car, they could die, be injured or be sued and jailed for homicide. "If a car slams on the brakes or turns in front of me and I can't brake in time and I have to go this way or that way to miss them, a lot of the times when I do that people shoot the bird at me because I had to cut in front of them. I think anyone who gets a driver's license in this world should have to take some kind of course on driving with trucks." Jeff says that to fill the shortage of drivers, trucking companies are hiring indiscriminately and training irresponsibly, giving safe truck drivers bad reputations and putting dangerous drivers on the road. "Some trucking companies put these kids through their school, then they get into a truck and stay with a trainer for a month. Once they get away from that trainer, after one month they let them be a trainer. You can't do that. Even one year ain't enough. That's why drivers are out here dying every day, because companies don't care. As long as they make the money, they don't care. To put kids in trucks and make them think they are truck drivers is terrible because they're costing other drivers their lives."

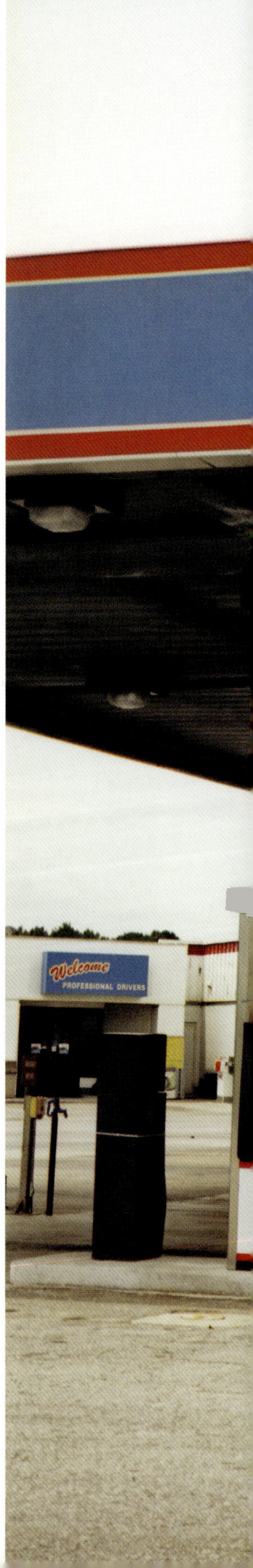

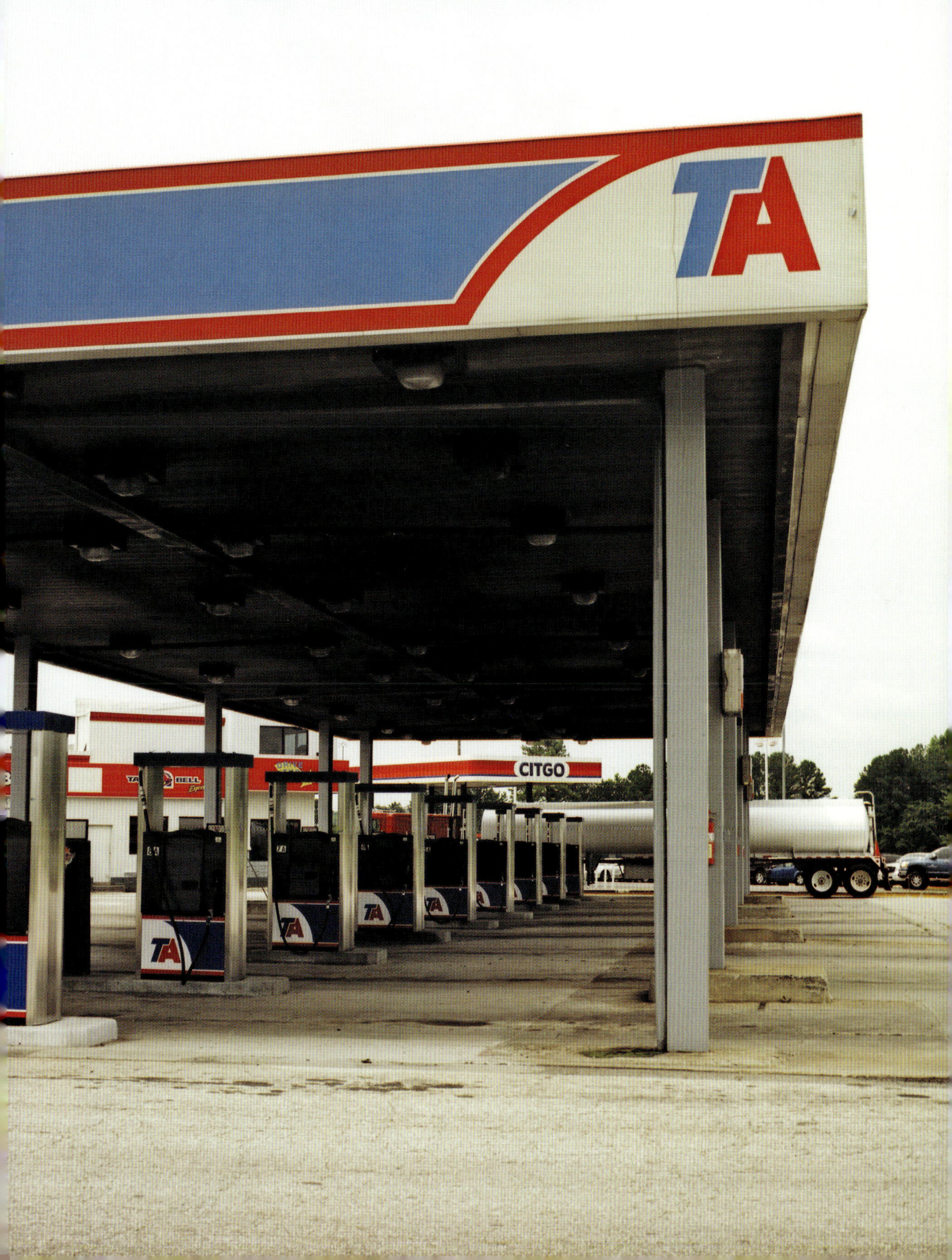
TA
CITGO

Bear

He's been watching us walk across the parking lot; when we get close enough, he says, "Original or Bar-B-Que?" then reaches back and gives us two bags of chips. He reaches back again and returns with a fluffy white dog named Bear. He is holding him so close to his face it is hard to tell where his beard ends and Bear begins. His beard, if shaved, could be used to recreate an accurate likeness of Bear. They are companions, Bear and the beard. When he sleeps at night, Bear lies next to it on his chest; if he sleeps on his side, Bear lies curled up under wherever it happens to fall. He is on the road for three to four months at a time. He has been doing this since he was 18. He says it was never intentional, just a flighty decision made nearly 40 years ago. "At a yard I worked at, a man was sick, so they asked me to fill in. I went in on Tuesday, I was in a truck on Thursday."

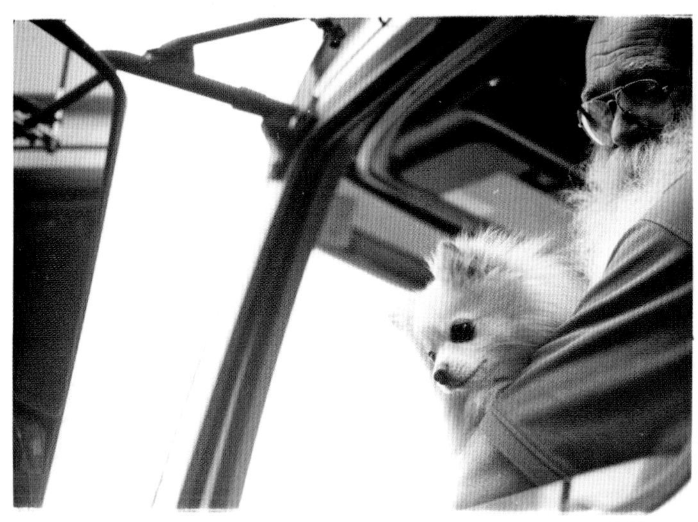

Jeff Ward; I-85, Exit 42

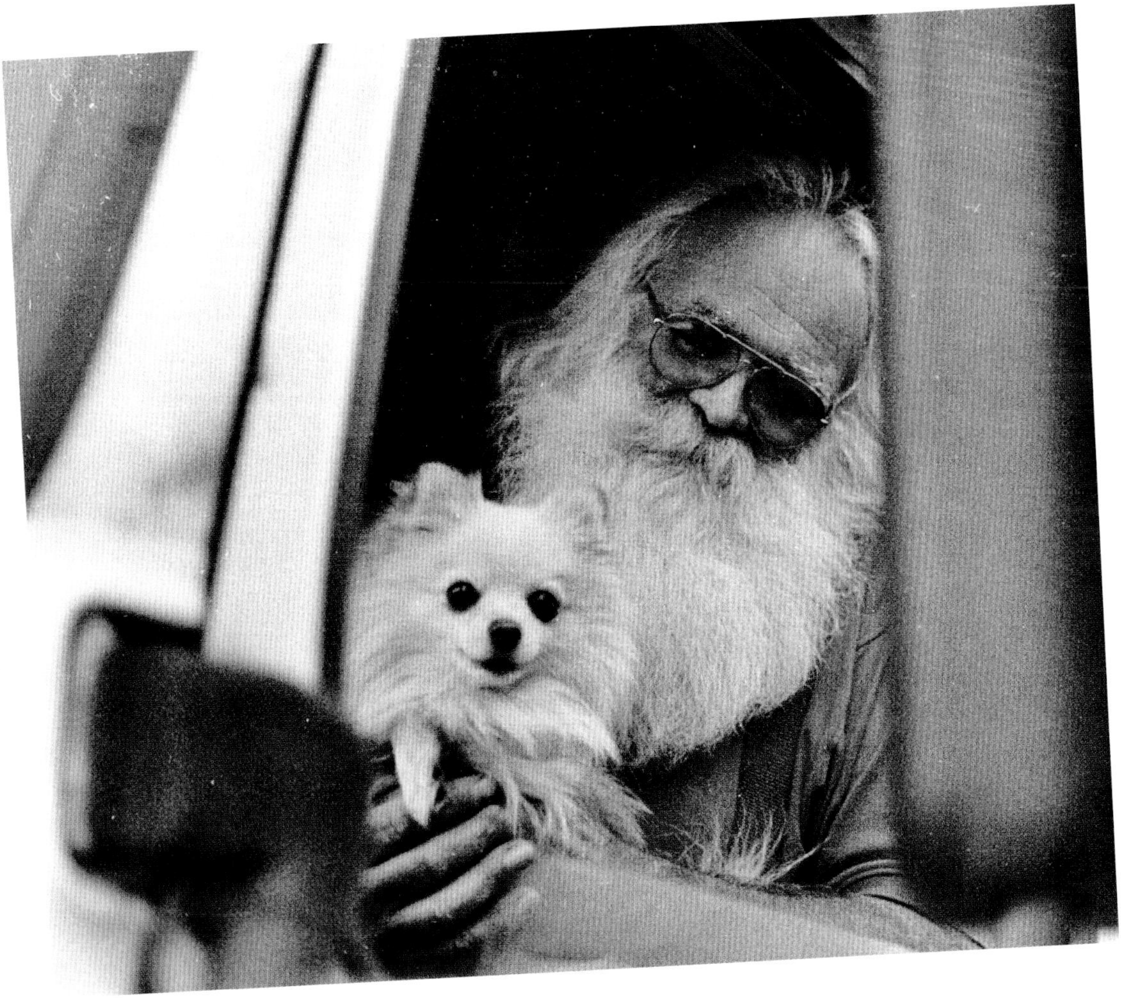

Commercial Harvest

Red agricultural trailers are loaded full of green bell peppers by a self-propelled harvesting machine manned by 105 pickers. The ripe green smell of the 150-acre field barely penetrates senses too distracted by wind, heat, dust and the noise of trucks and tractors. Brown hands place the peppers on a conveyer belt that carries them over the edge of the trailer. A tractor pulls the trailers into a dirt lot next to the field to be attached to a truck. Robert Duarte, the truckers' field supervisor, says once they arrive at Con-Agra, half will be sliced for Subway and the rest for Hometown Buffet. Four fallen peppers are smashed while three truckers tarp the load. Behind them, more trailers are being loaded; Robert says the drivers come here from Mexico for harvest season. "It begins in Imperial Valley in early May and continues for six months, with the truckers and the pickers moving north, following the harvest." The drivers are paid $130 per load. Con-Agra hires a grower, who hires planters, pickers and truckers. The truckers live in Modesto, California, near Con-Agra. They drop the peppers and wait to be called back the next day. Taco, Con-Agra's overseer, began his career 44 years ago as a 13-year-old picker: "The tomatoes last week were for McDonald's breakfast burritos. It was all the same workers." Today it is 103 degrees. "This is nice weather. Last week it was 117 and a lot of them were all passing out from heat stroke."

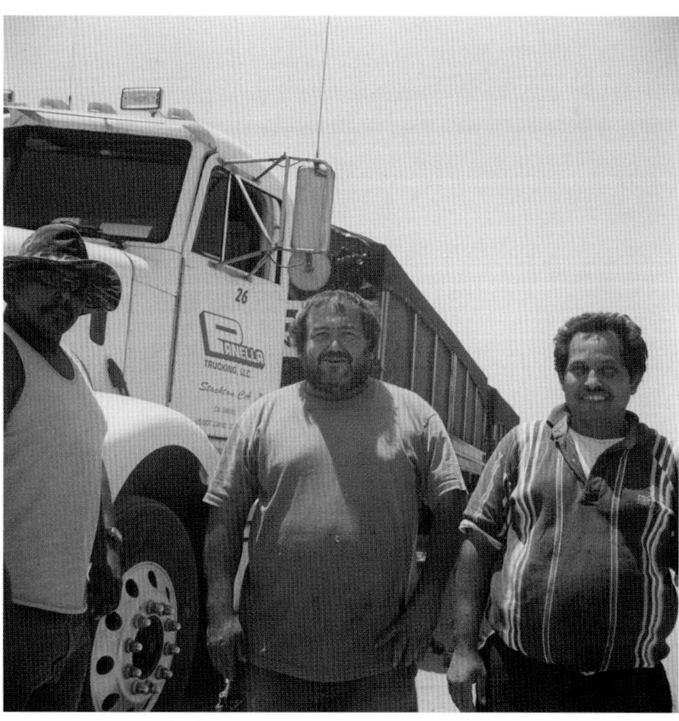

Harvest truckers; Hwy. 99 and 7th Standard Rd. Bakersfield, Califorinia

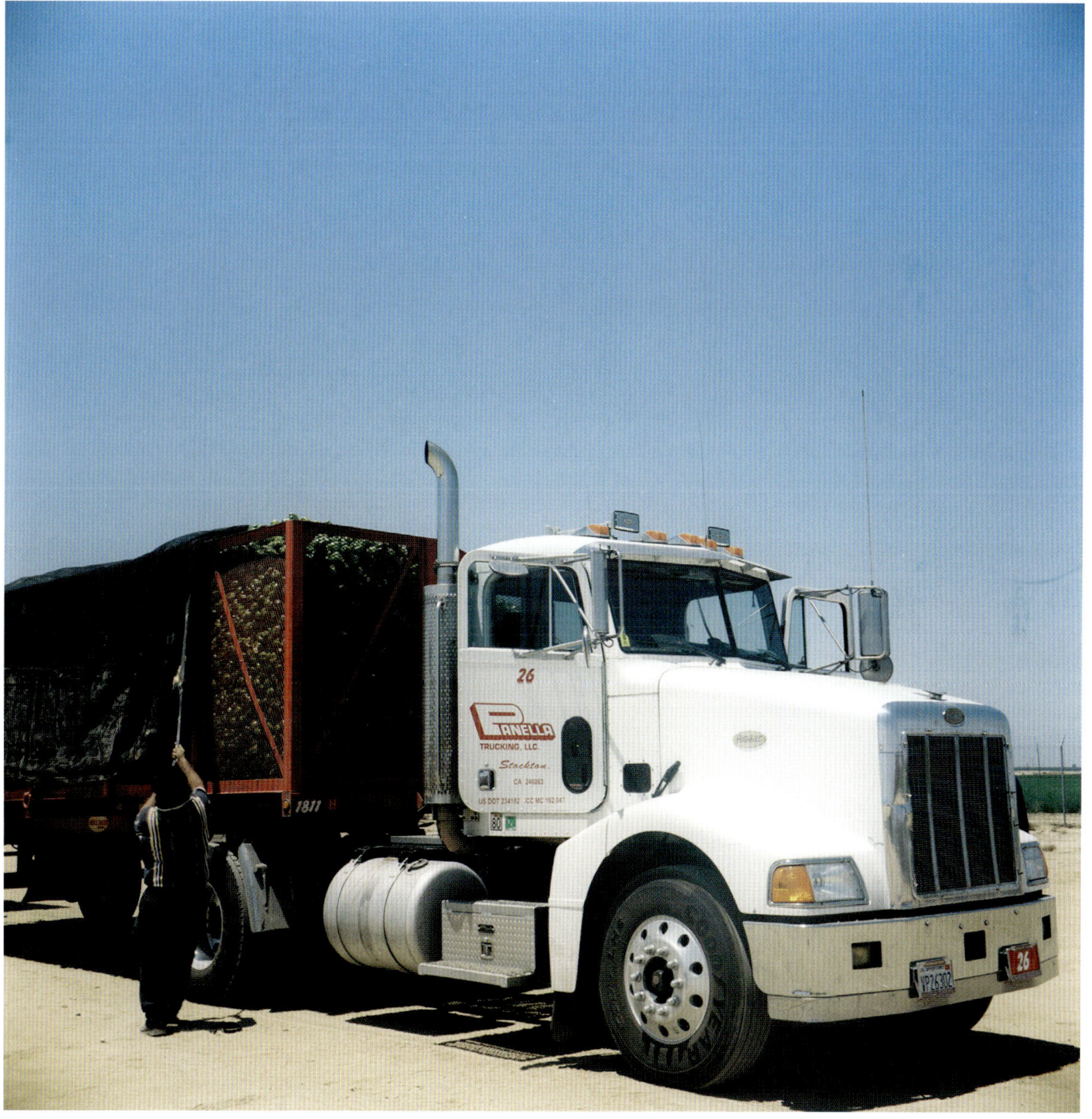

Tarping a load of green bell peppers

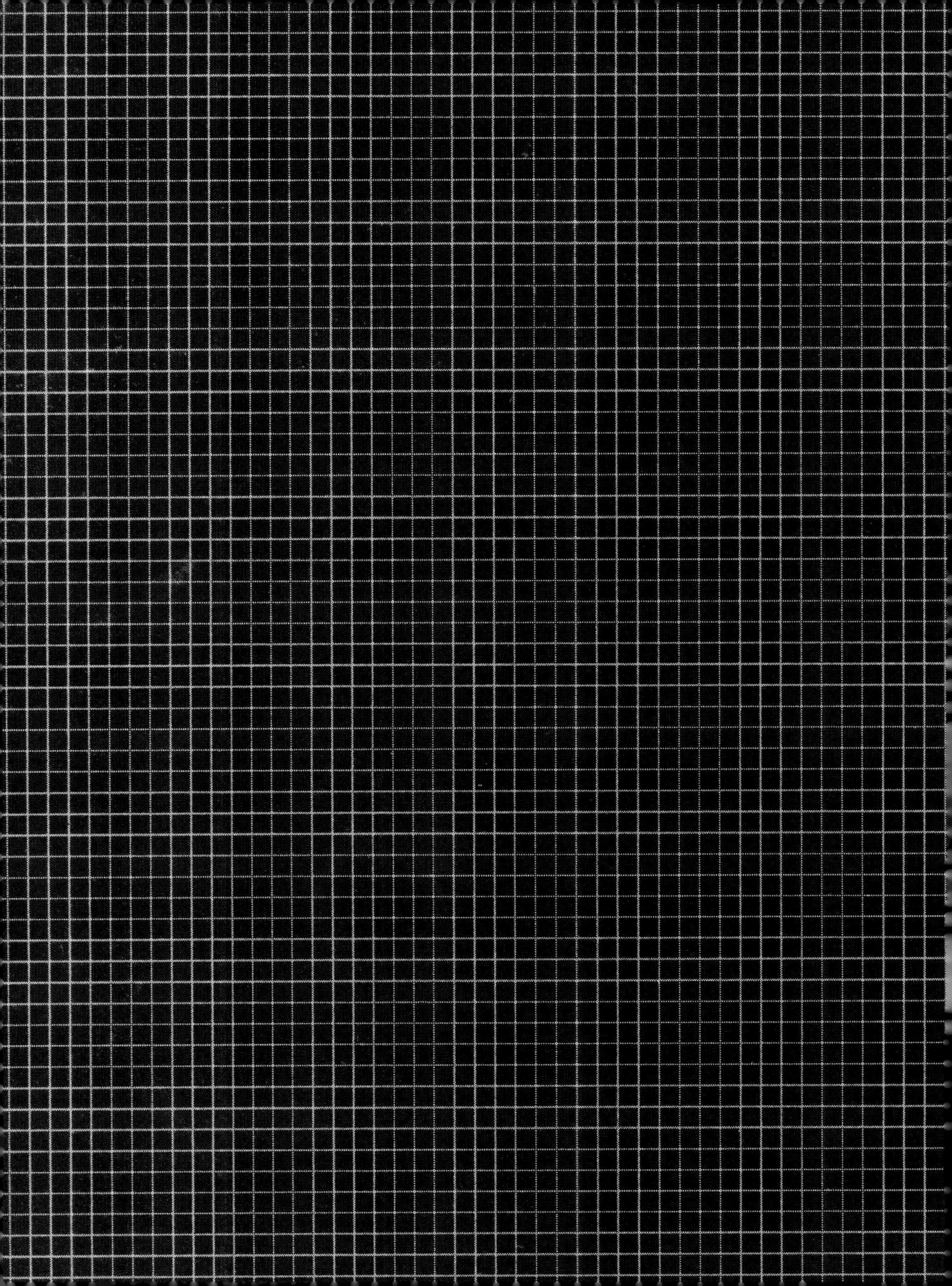

PART TWO

Passing time, everyday life, needs

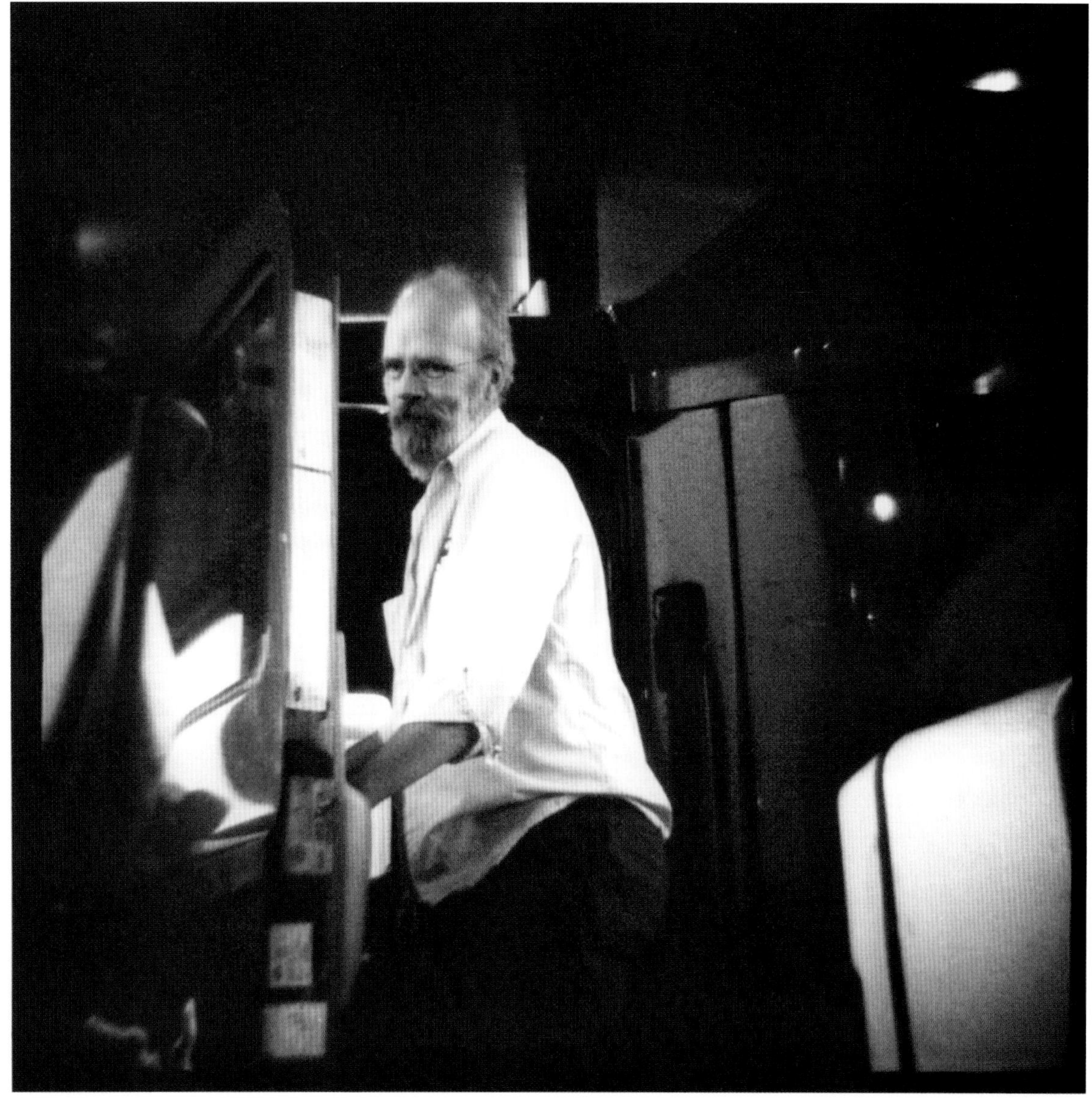

Rob, hauling Kaoline, a clay used in manufacturing glossy paper

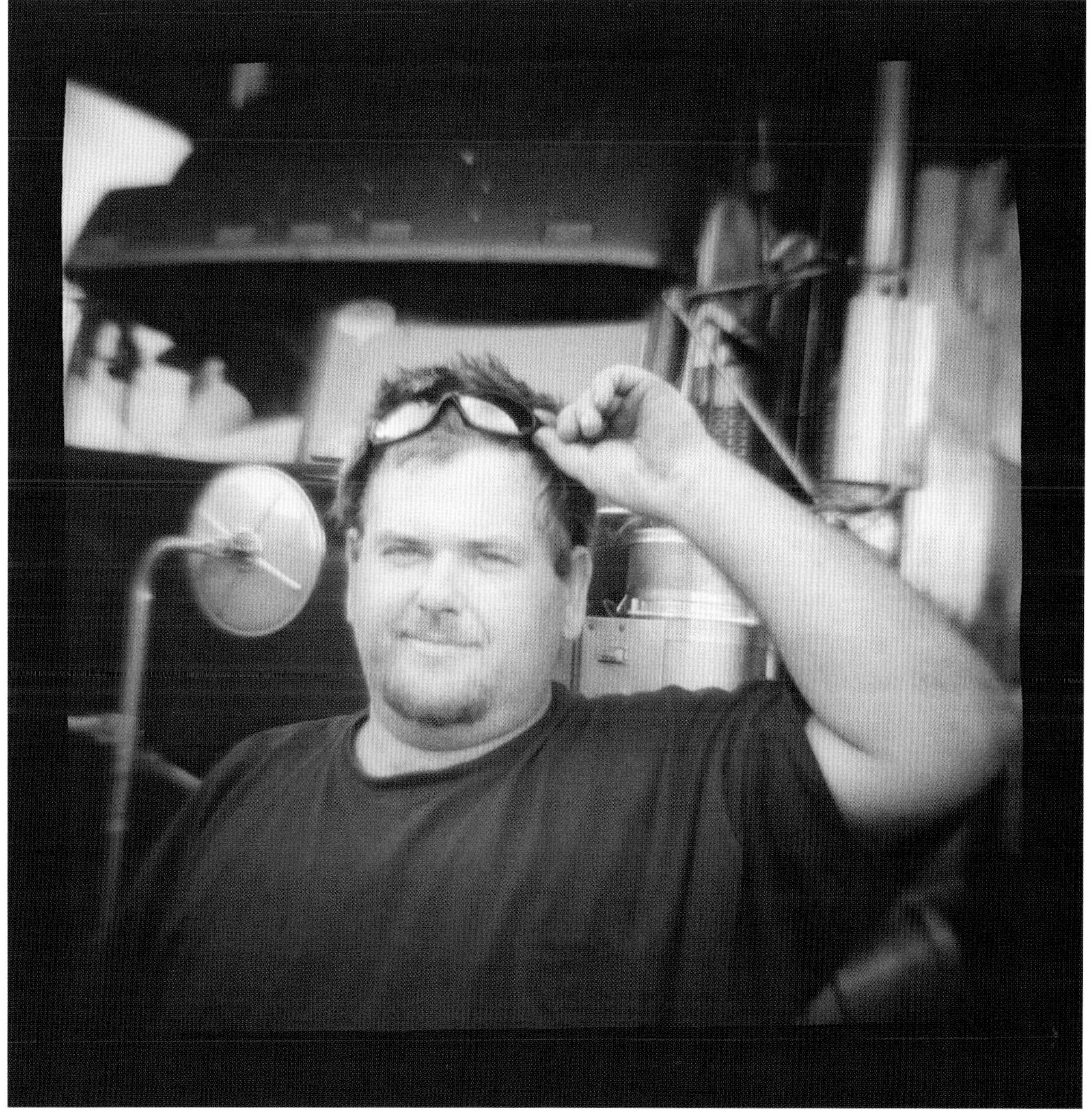

Joseph Case; I-85, Exit 129

Charles Tesar

"We're an oil-based society. People don't understand this. It ain't the gasoline. Fuck the gasoline. Only 50 percent goes to gasoline. When we don't have these supplies, we'll go back to the cavemen."

Trucks run on diesel. Charles Tesar lives in Brooklyn, New York. He's sixty and has been driving for about 40 years. "Years ago, the trucks brought it to a warehouse. We don't do that anymore. That's why they changed it, to make more money with it. It's going straight to the stores to save money and to make money. One time, trucking was a service from point A to point B, safely. You could allow on your course time for the maintenance of your equipment. Now it's production." He used to work on his own trucks and drive across the country. But age and technology have put new limits on him. "I'm just a driver now. I used to have my own tractor. I gave it up because of the computer. At one time, they were mechanical. You could take a screwdriver and fix it." The design of the truck and the industry has evolved, but the nature of the people drawn to this work is unchanged. "To be a long distance trucker, you got to be a wanderer. That's how I used to be. Now I'm too old to be wandering around, so I stay local. Today I'm a bum. Yesterday, I was a hero."

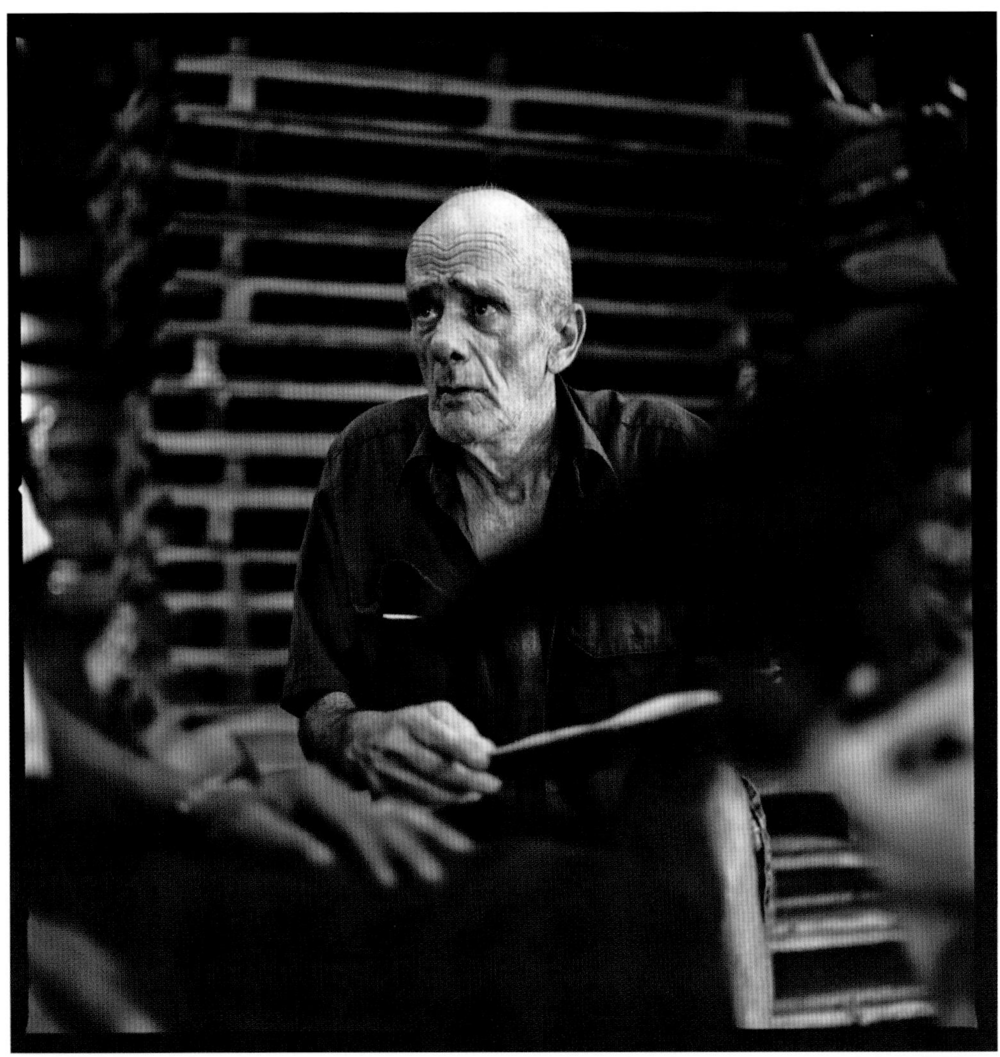

Charles Tesar, receiving paperwork

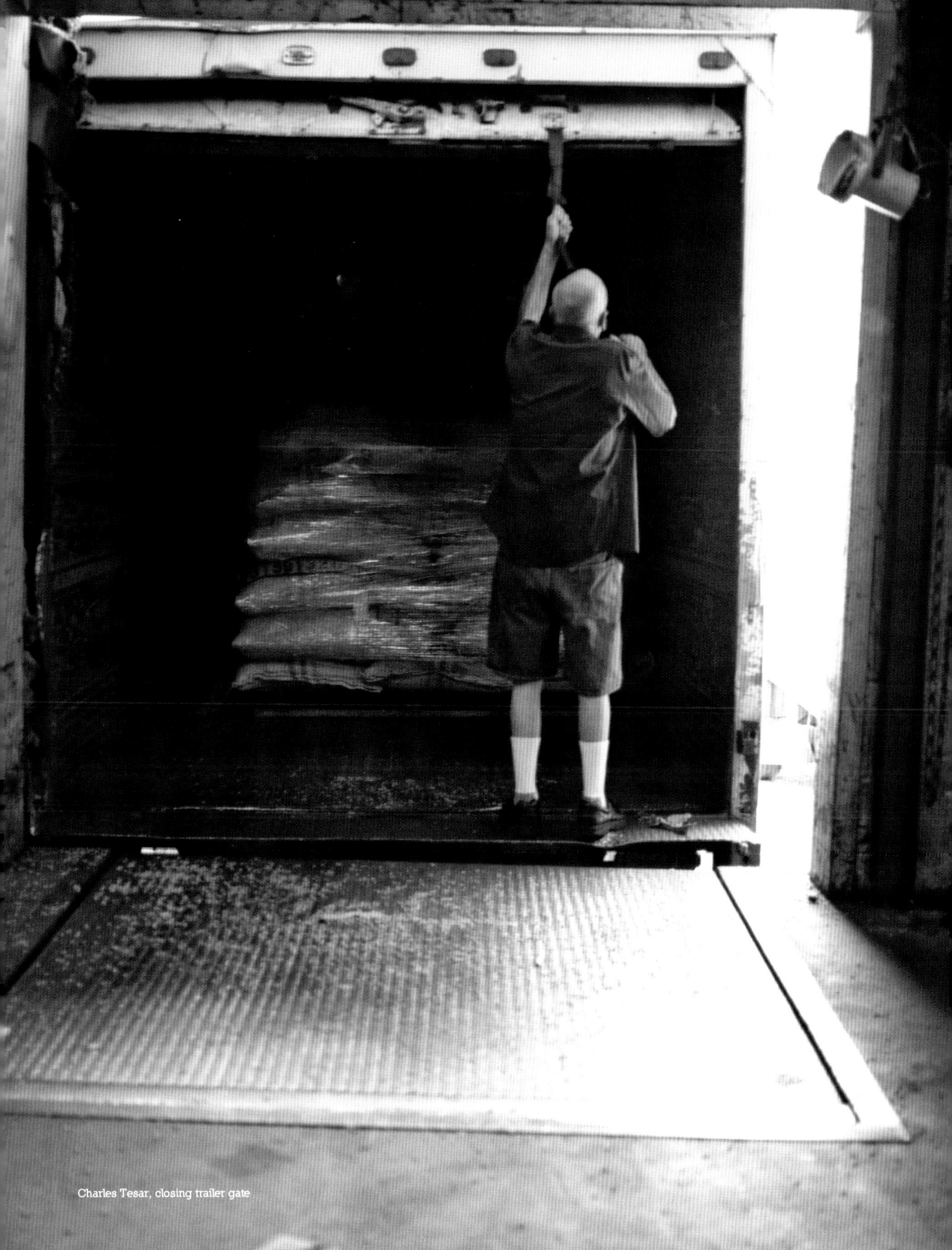

Charles Tesar, closing trailer gate

Donna's

"The big hot tub is down and the neon sign was broken during the earthquake," Gayle says regretfully. It has been a slow year. "In 2000, this block was real packed. Over a hundred trucks used to stop here. Once a week, I make what I used to make in a night."

Gayle is the bartender and she leans over the counter, smoking and watching the door. One driver is here to use the free shower and free internet. Amy, one of the residents, comes into the bar wearing heels and a red corset. She walks over to him and leans over his shoulder while he's checking email: "You are a very sanitized man." He does not respond warmly; she sits for a while, then returns to her room.

When it is busy, Gayle counts clicks on the ATM, sets and monitors the timers, watches for trouble, does the books and cooks for the house. For truckers, there is free chili in a crock-pot near the door; for the girls, crab salad in the kitchen. At one time, Gayle was married to a trucker for seven years. "I did the logbooks, I did the paperwork, I was left at home with young kids. It's a hard life. They come home cranky and tired and so are you. He had two trucks and lost them. He was doing good for a while, but buying the second one is where he went wrong. Back then it was a good living, Now it's not."

Business at Donna's is directly affected by the trucking industry. The girls take 30-minute shifts on the CB every day from 2 p.m. until midnight to advertise to drivers, who make up around 80 percent of their customers. "We aren't as busy now because the truckers don't have as much money. We're happy if we get one shift that gets ten." On the wall behind her are handwritten signs announcing how to access the free WiFi and a list of souvenirs: semi mud flaps, screw drivers, ice scraper, tape measures, maps, flashlights, caps, mugs. The main hallway is lined with narrow wooden shelves filled with Donna's coffee mugs signed by truckers: *Ragin Bull, Coffee Pot, Suicidal Maniac, Speedy Bruce, Mallard, Krazy Horse, Booggy Booggy, Gator, Buzz*.... Another casualty of the recent earthquake, the four lowest shelves are empty.

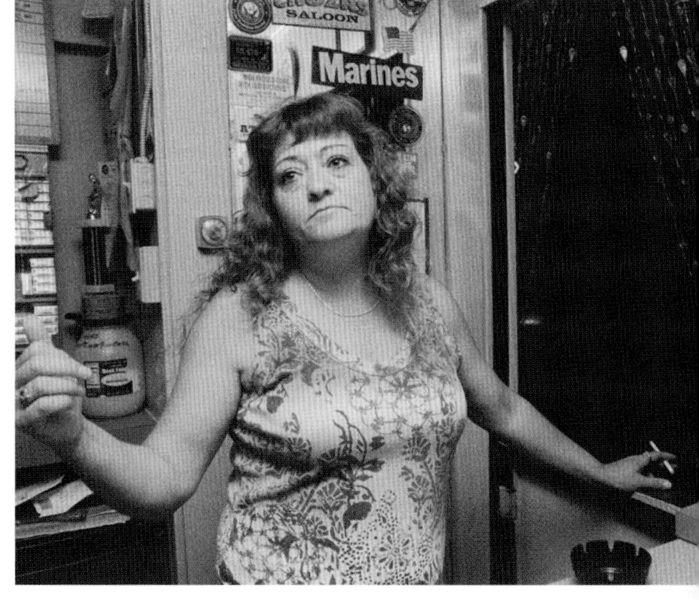

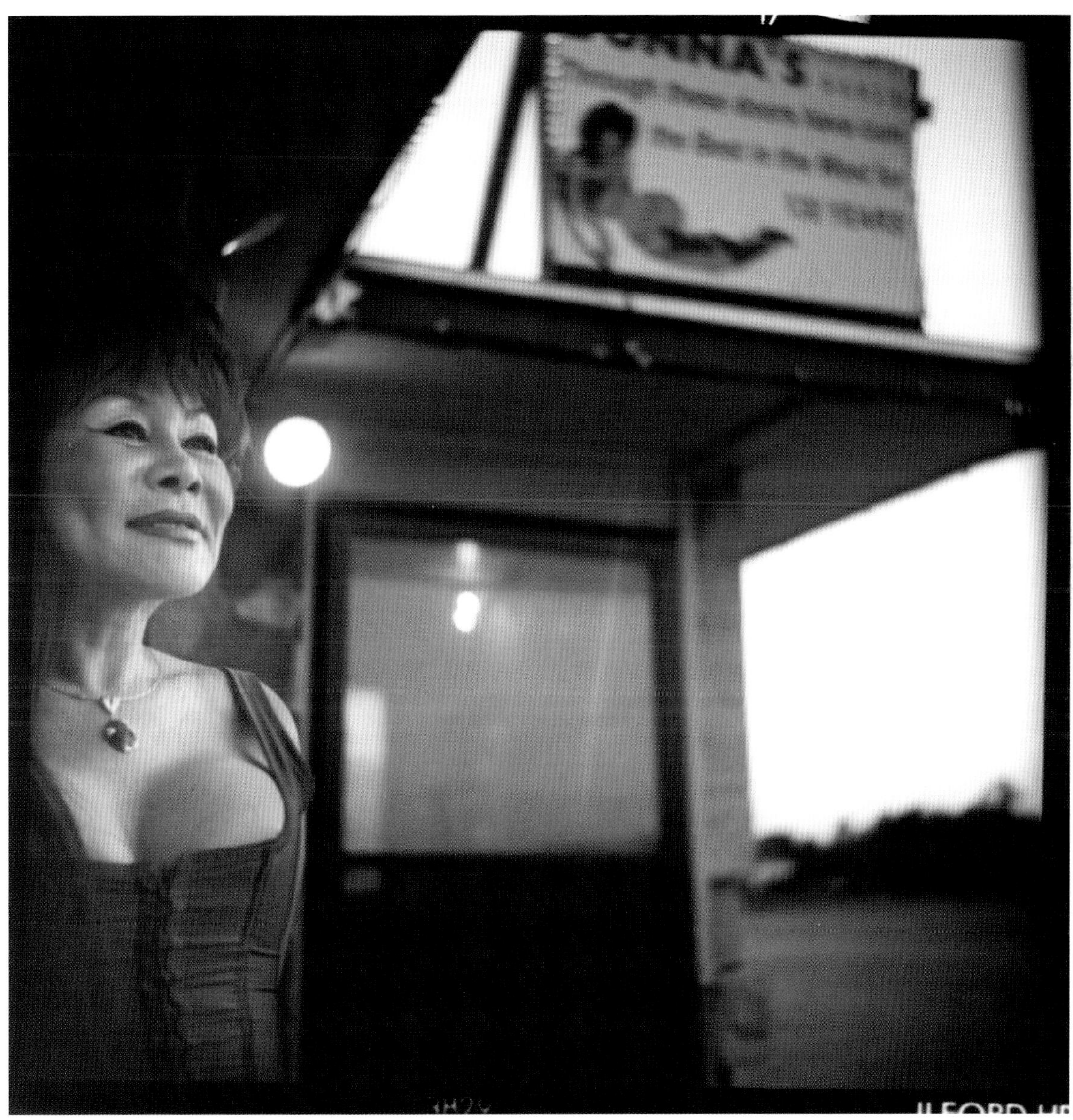

Gayle (previous page) and Amy;
Donna's Ranch

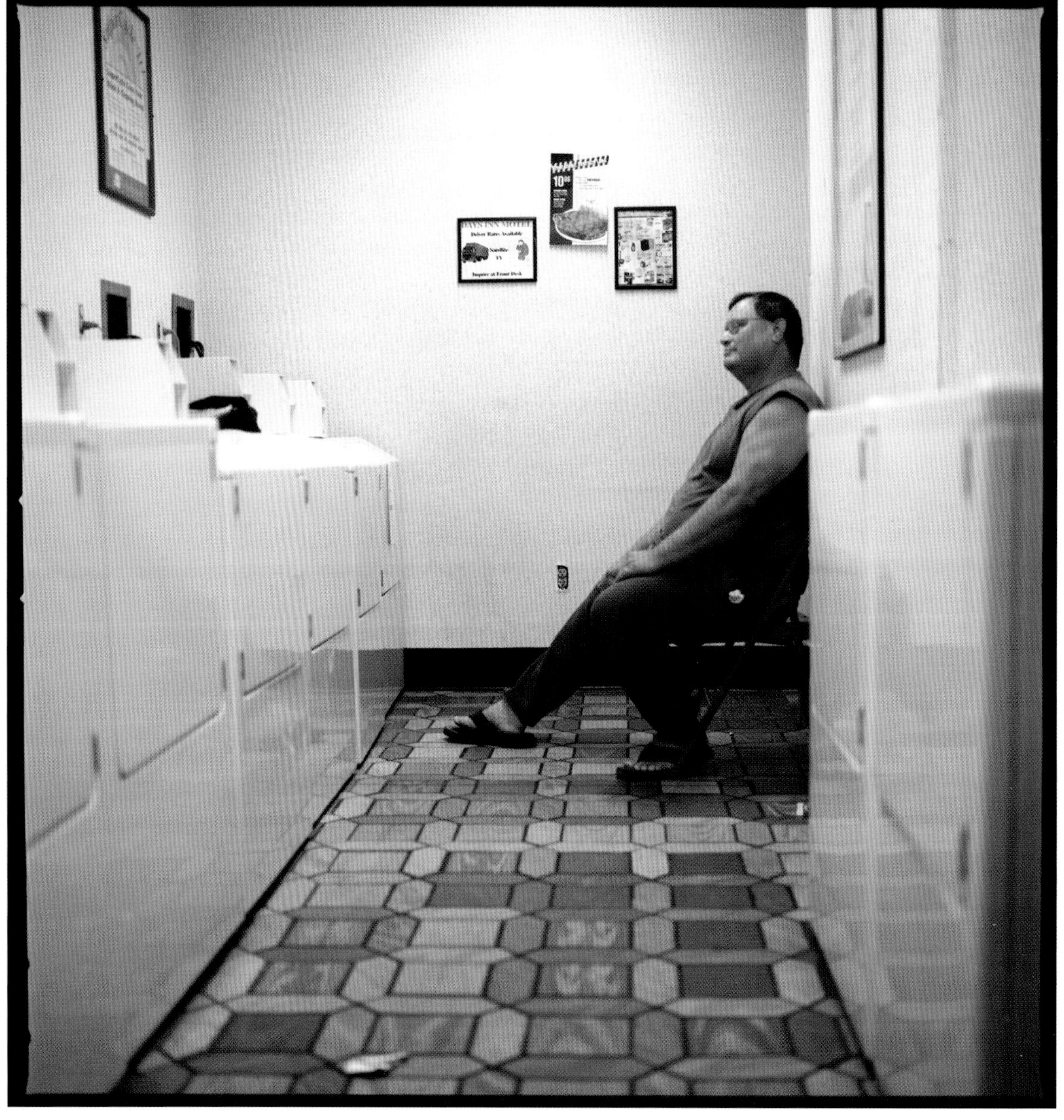

Craig Worley washing laundry; I-80, Exit 4

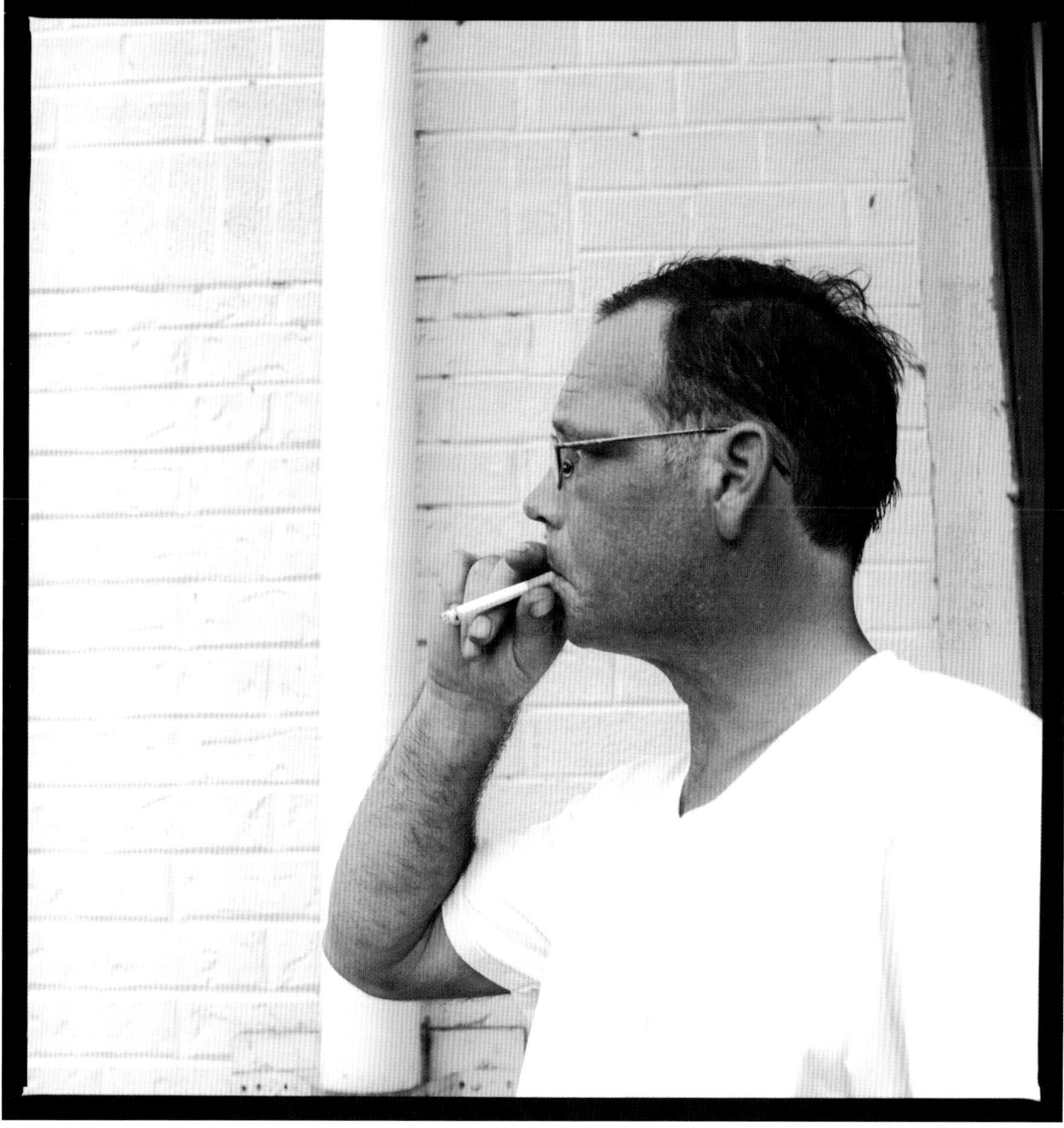

Floyd Mallory, smoking outside of a truck stop; I-80, Exit 4

A New Life

Howard Capeles spends three or four months out at a time. "I'm not like some drivers who spend their break driving home." He is training another trucker; the two will ride together for 30 days. They are backed up to a small airport and planes are flying overhead. They will be parked here all weekend until the load can be delivered on Monday. Thirteen years ago, Howard chose to drive a truck because he was unhappy with cheap factory work and fast food restaurants. He is 41 with a teenage son and a teenage daughter. His trainee is quiet and tired; the skin on his face is sunken. He is not here. Quietly listening to the planes, he drifts further into a future without Howard. Clouds shift over the sun and a resigned acceptance shadows him, a slipping on of his new identity.

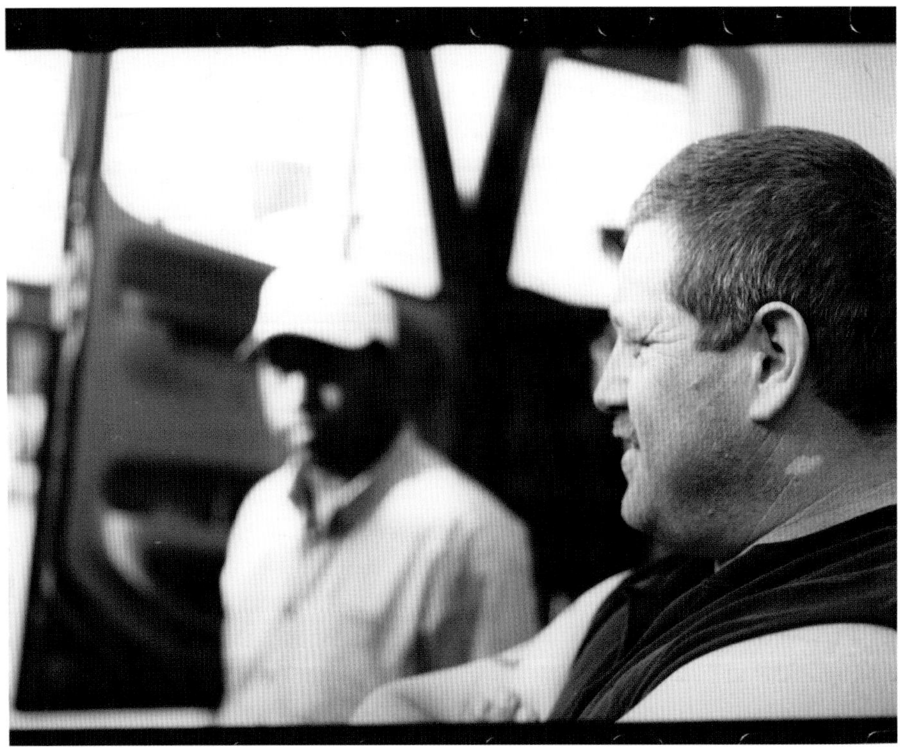

Howard Capeles and trainee; I-85, Exit 42

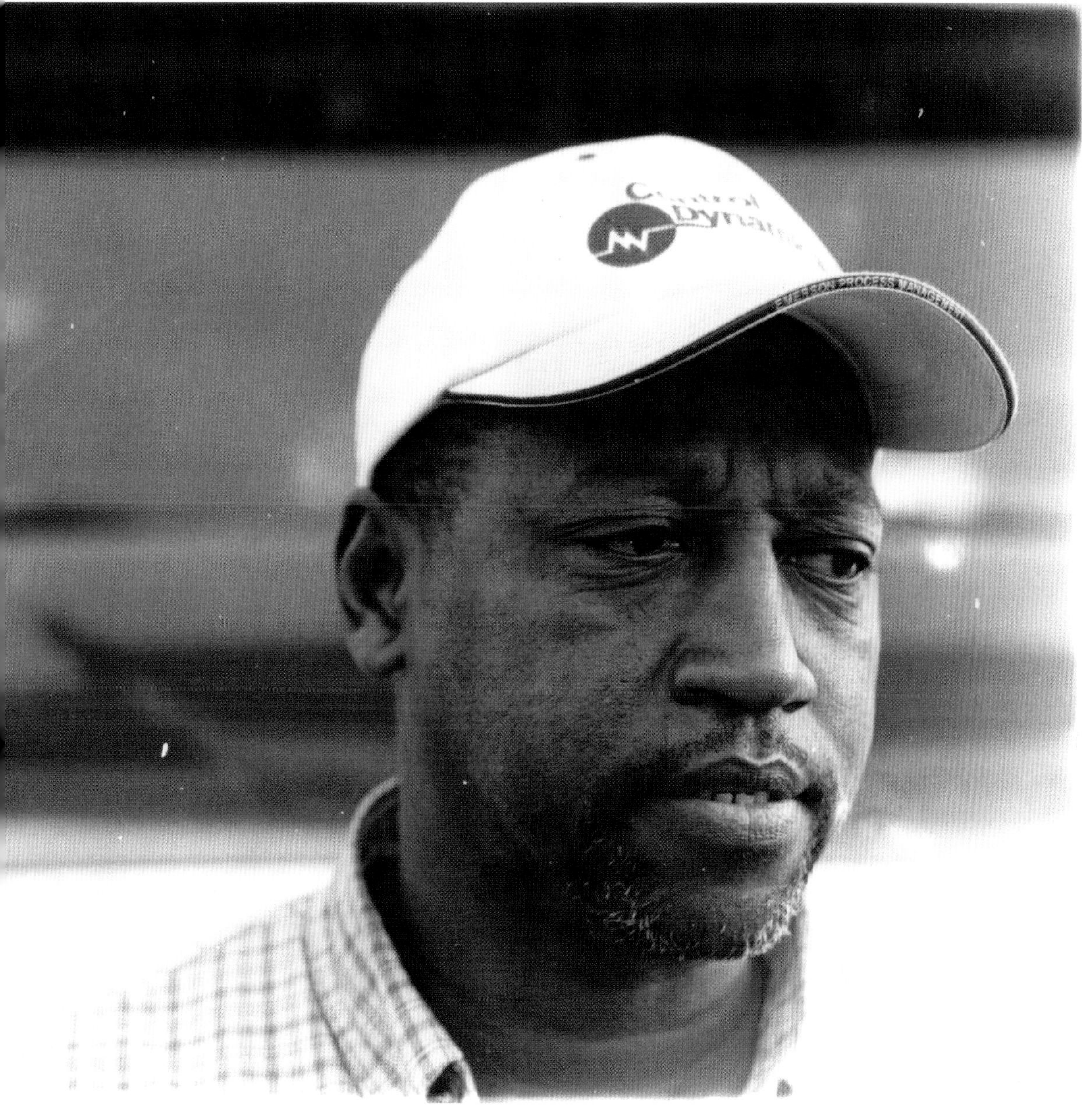
EMERSON PROCESS MANAGEMENT

Exit Does Not Exist

It is noon on a hot, windy Sunday and dust whips around our faces. Bob Hutchinson is parked in a near-empty dirt lot for the rest of the day and part of tomorrow with his load of Kawasaki lawn mowers. His truck is loud in its idling. Bob located this place in his *2008 Truck Stop Directory*. "I plan about two hours out, thinking, Where I'm going to lay down for the night? I start looking through there. I'm on 16. I've got to get off at mile marker 143, so this is the closest one to me. So it says it's a medium facility that will hold up to 50 to 100 trucks. It's supposed to have a restaurant, a truckers store, country store, showers, laundry, truckers lounge and truck repair. But when I pulled in last night, I thought it was closed down. I had 15 minutes left to get wherever I wanted to go, I'm freaking out, thinking this place is closed down, and I pull in and look at the door and the sign says, 'Be back at eight.'"

Truckers can work no more than 70 hours in an eight-day period, no more than 14 hours a day. "Whenever you start, your 14 hours start running. You use 11 of that driving. You can split it however you want it, but at the end of 14th hour, you're done. At the end of eight days, you can't have more than 70 hours. You can drive eight and three-quarter hours a day and get eight and three-quarters back, enough to go along. But this won't be delivered 'til Monday, so I ran my ass off from Lincoln to here and now I'll sit for 34 hours so I can get my full 70 back. It just depends on how hard you want to run."

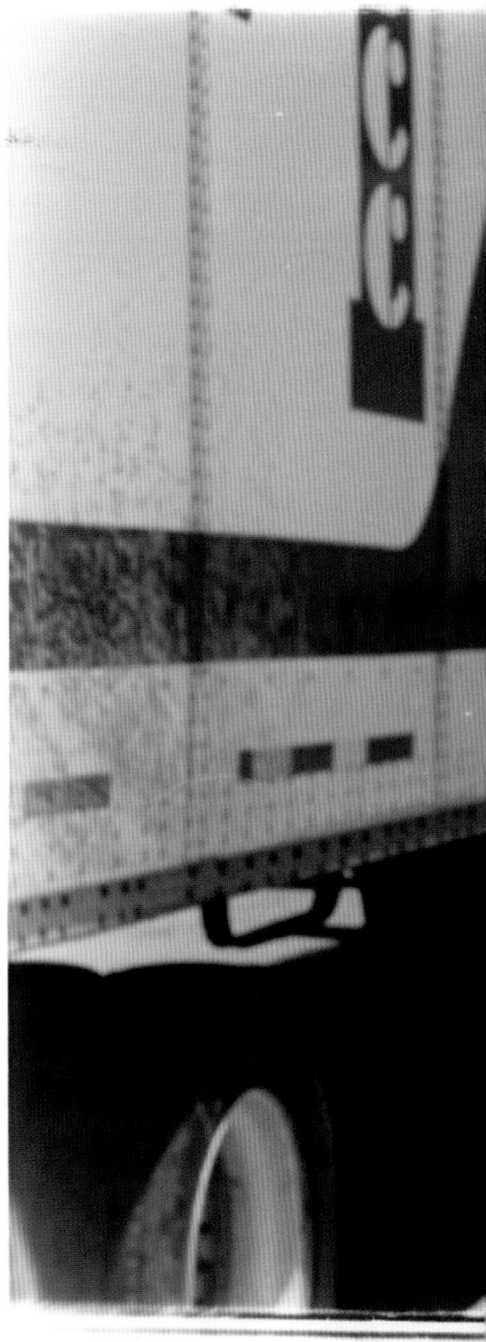

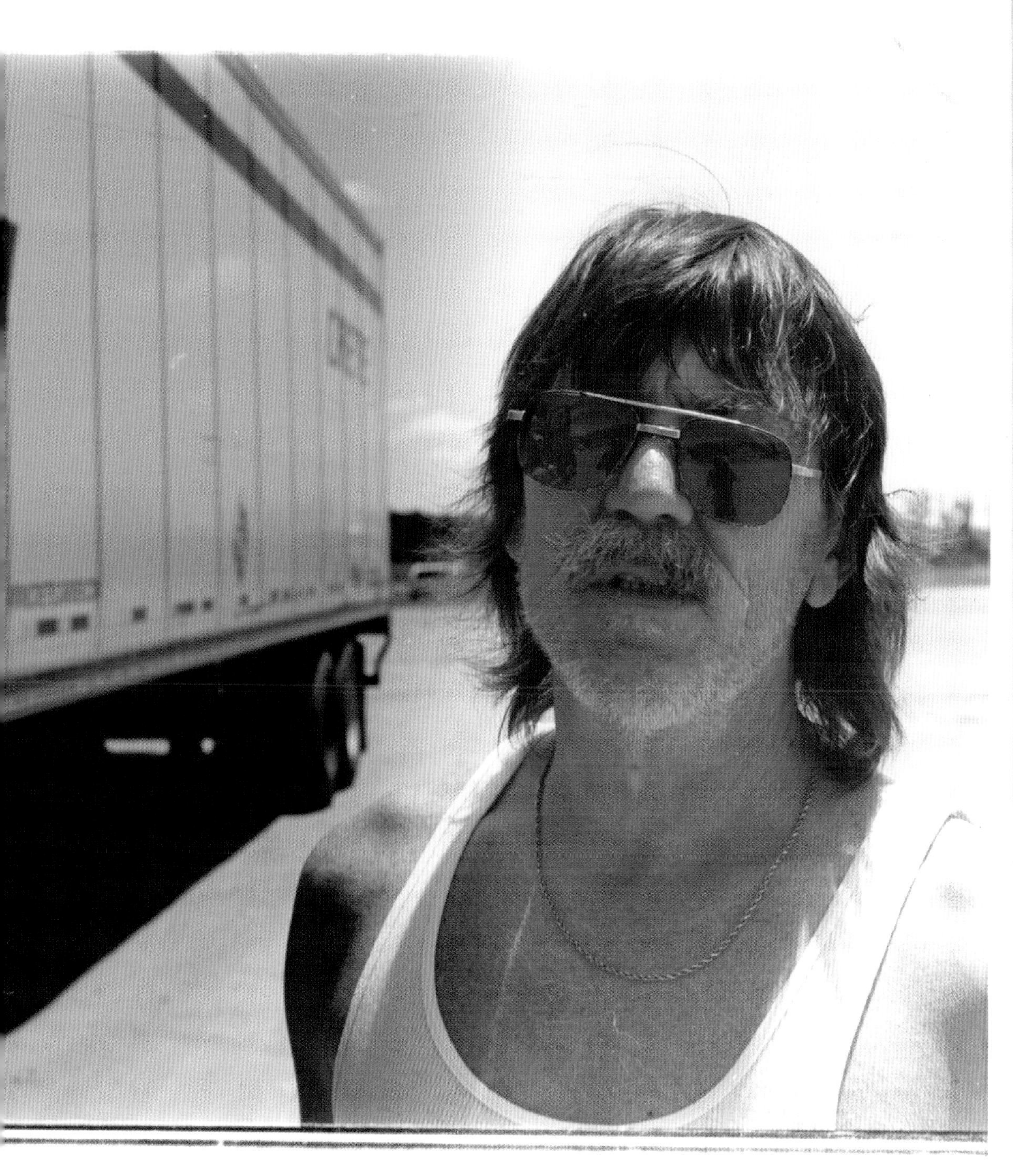

Bob Hutchinson; I-16, Exit 111

Suicide

"A truckload of cigarettes is worth over a million dollars," says Dave. He used to run from Phillip Morris in Richmond, Virginia, to California before becoming a flatbed driver. He says he was escorted by police upon entering Los Angeles. Drivers of high-value loads like electronics, DVDs, copper coils, cell phone batteries, usually don't know what they are carrying unless it is essential to the delivery. On a flatbed trailer, high value loads are tarped so that the load is not visible, like copper coils. Tarps weigh 120 pounds and must be secured by the truck driver. In the winter, tarps might become frozen solid. A trucker can lay it on the ground and drive over it to let the engine thaw it out. Dave points to the back of the cab of the truck. "A headache rack is supposed to stop a load on a flatbed trailer from crushing the driver in an accident. But it would just go right through." Some trucking companies put stickers on the side of trucks that are driven by team drivers to indicate to paramedics that there are two drivers. This will increase the chances that they would search for a second driver in an accident. Team drivers can ship loads twice as fast. One person sleeps while the other drives. For high-value loads, it is sometimes essential to have team drivers. Companies will load a flatbed for the driver, but after that, he is responsible for securing it so that it does not become loose and fall off while driving. "If those rolls come off and kill somebody, my life's gone. It's my responsibility. I basically murdered that person." Truckers thus learn different ways to secure loads. Industrial coils are a typical load, and there are three ways to stack them.

1. Flat on its side: "eye to the sky"
2. Sideways: "shotgun"
3. Positioned so that they roll front to back: "suicide"

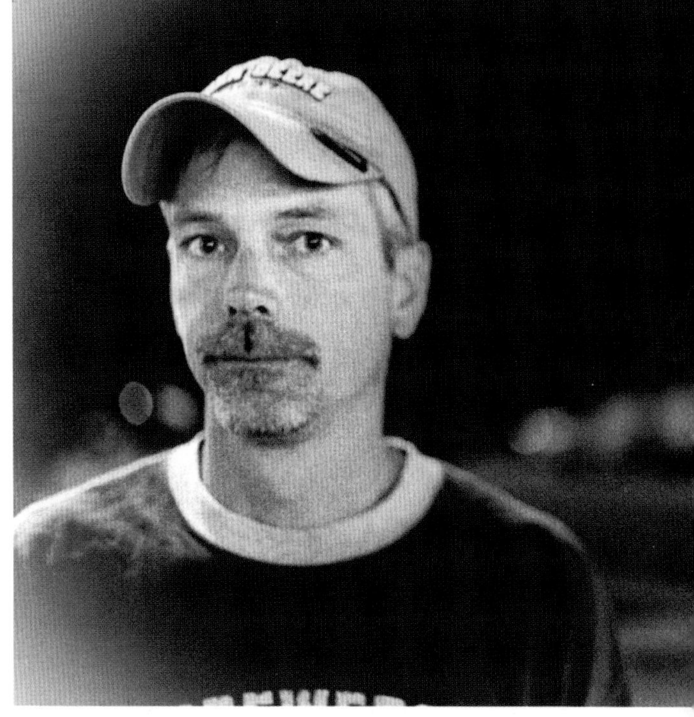

Dave; I-85, Exit 160

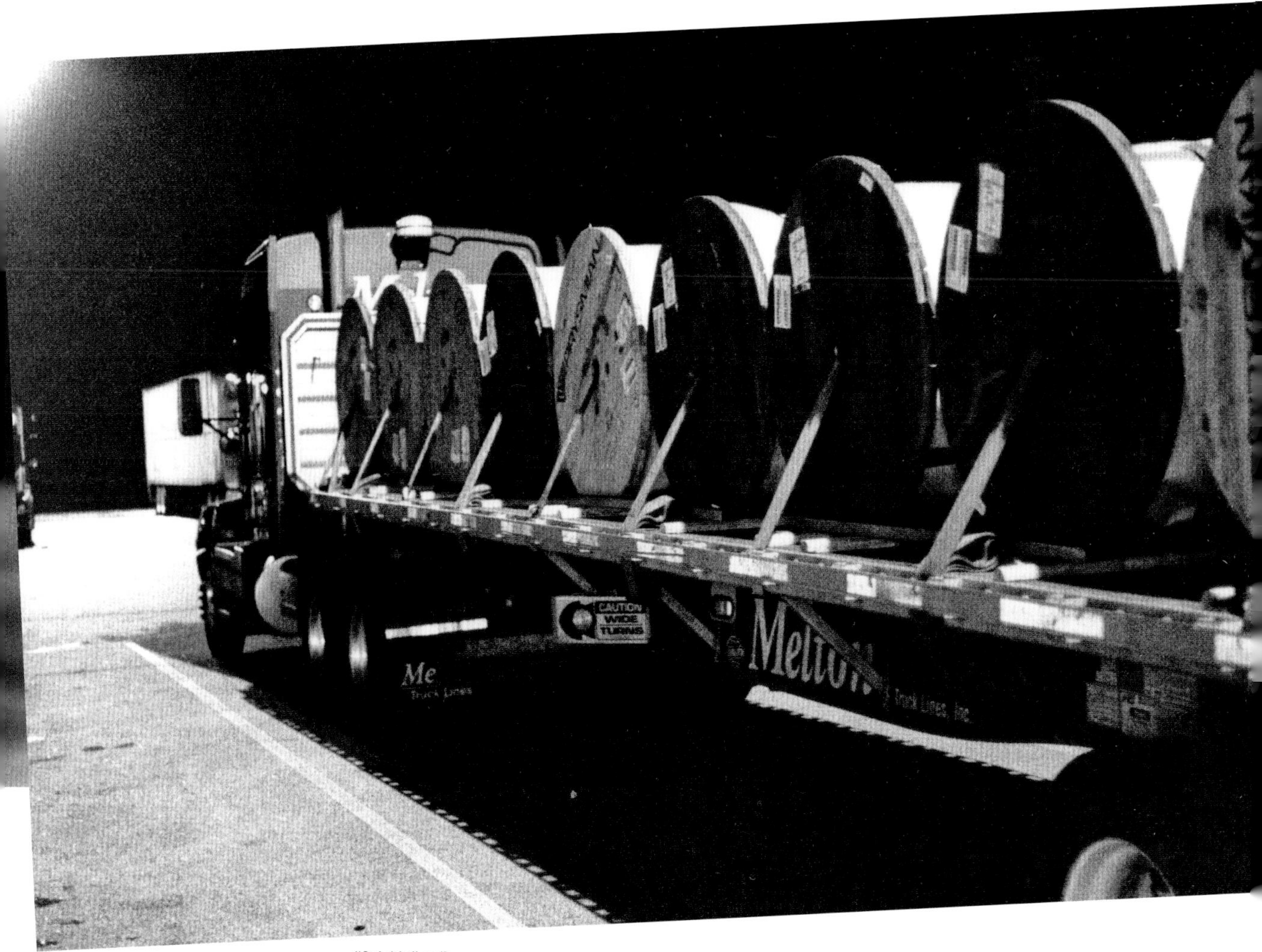

"Suicide" coils

Truck Payment

Michael Flowers hasn't had a paycheck in nine weeks. He has a lease purchase plan with his employer. He pays $291 a week for his truck, which is taken from his check. "It's a 36-month lease. I'm into it eight-and-a-half. There's a $2,100 balloon note at the end. Every month I pay for my truck successfully, the company kicks in $200."

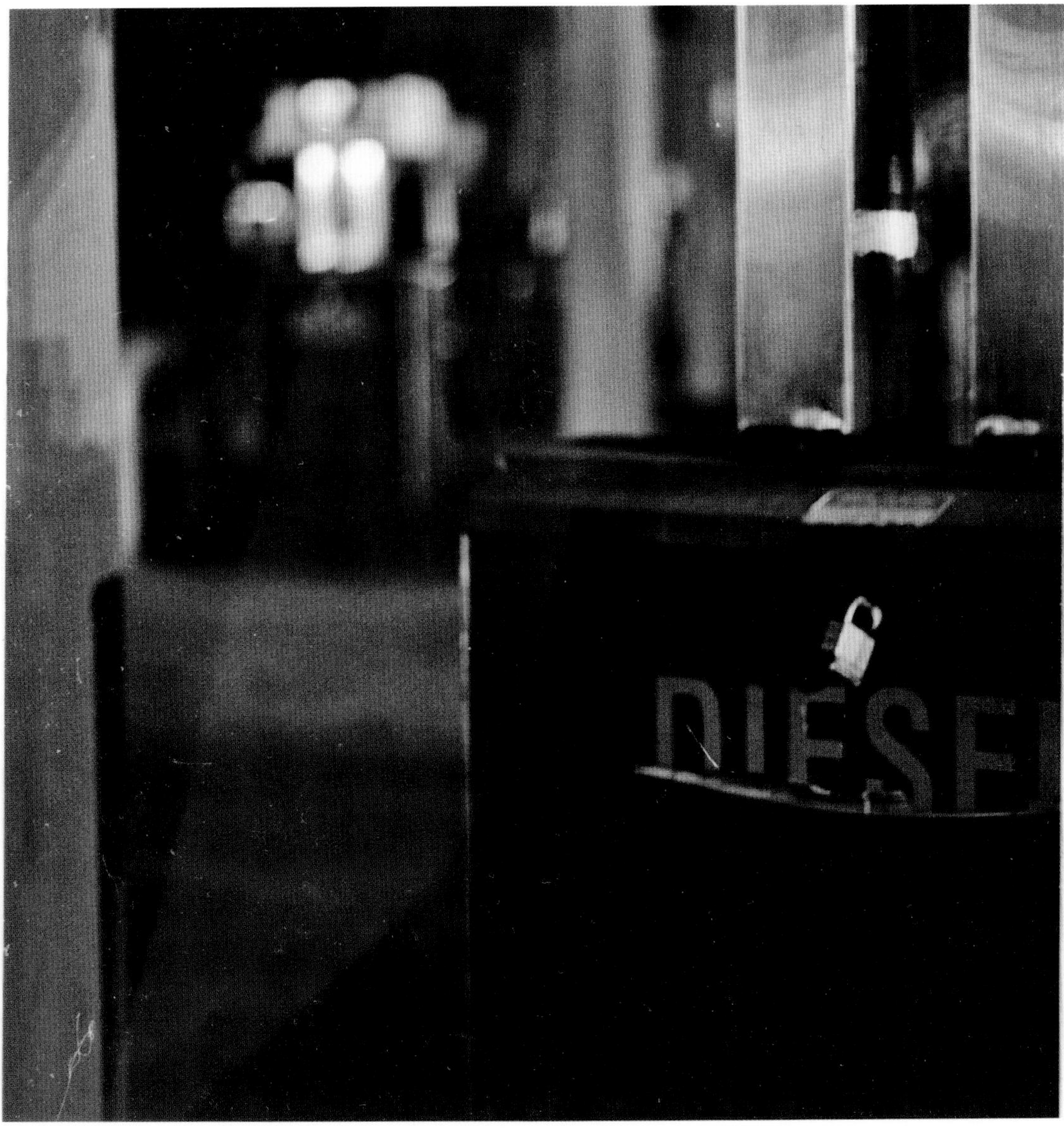

257 Truck Plaza

The smell of cow manure drifts over from a nearby field. One truck is parked in the grass growing beyond the gravel. A cement island is fitted with four fuel pumps and two racks of rags and cleaning products, forming the epicenter of 257 Truck Plaza. It opened in 1971, but then it was called Barney's. The owner, Shekh Rahman, greets us with curiosity and a slight hope that we are interested in the property. "Things started to slow down two years ago when the Love's opened. I have other business, that's why I'm alive. But I'm not going nowhere unless someone writes me a big check so I can go someplace else and do something else." Inside 257,the detrimental Love's truck stop is in perfect view from the front door, a quarter of a mile away.

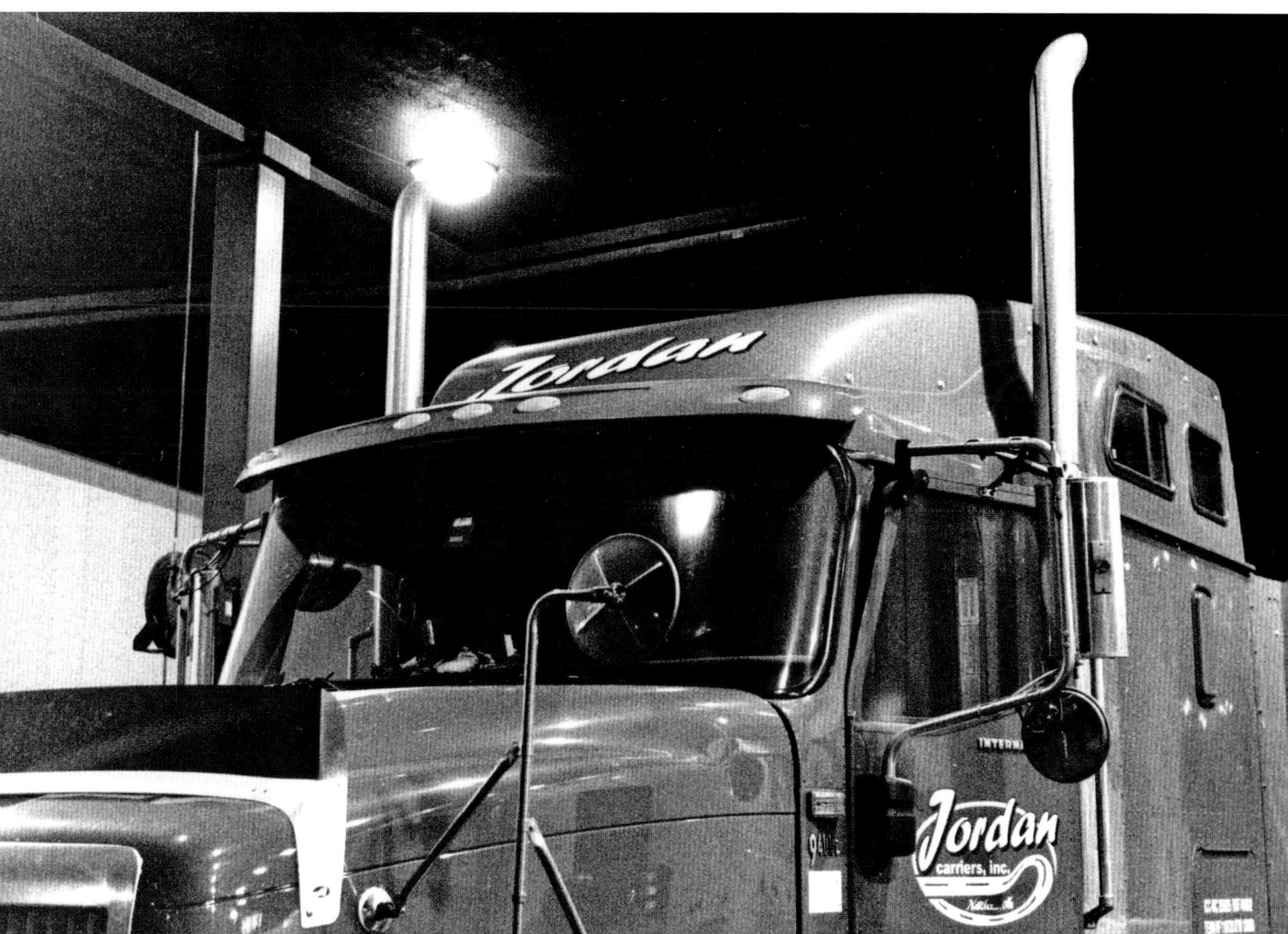

Parked truck at 257 Truck Plaza

On Speed

Lisa is standing behind a dumpster in a windbreaker. "This is the ghettoest truck stop there is. The truckers sell fuel, tires, anything. People in those apartments back there come out here and steal toilet paper out of the trucks." Her eyes are cloudy and bulging; her teeth have lost their shape, the edges worn by decay and neglect. "I smoke crack, I ain't gonna lie. I'm out here working." She throws down her cigarette to wave to a trucker who just pulled in. In two minutes, she's back at the dumpster. "I was left here three months ago," 588 miles from her hometown. She notices a new truck, runs out to it, climbs the step and leans into the window. It is not dark yet. The pavement is stained black with oil and there are puddles of water collecting in dips. The air is full of mist and lonely, busy tension. Most truckers turn her down. One sees her approaching and he furiously shakes both of his hands, pushing the air as if physically pushing her away, pushing at the edge of some invisible force field between them.

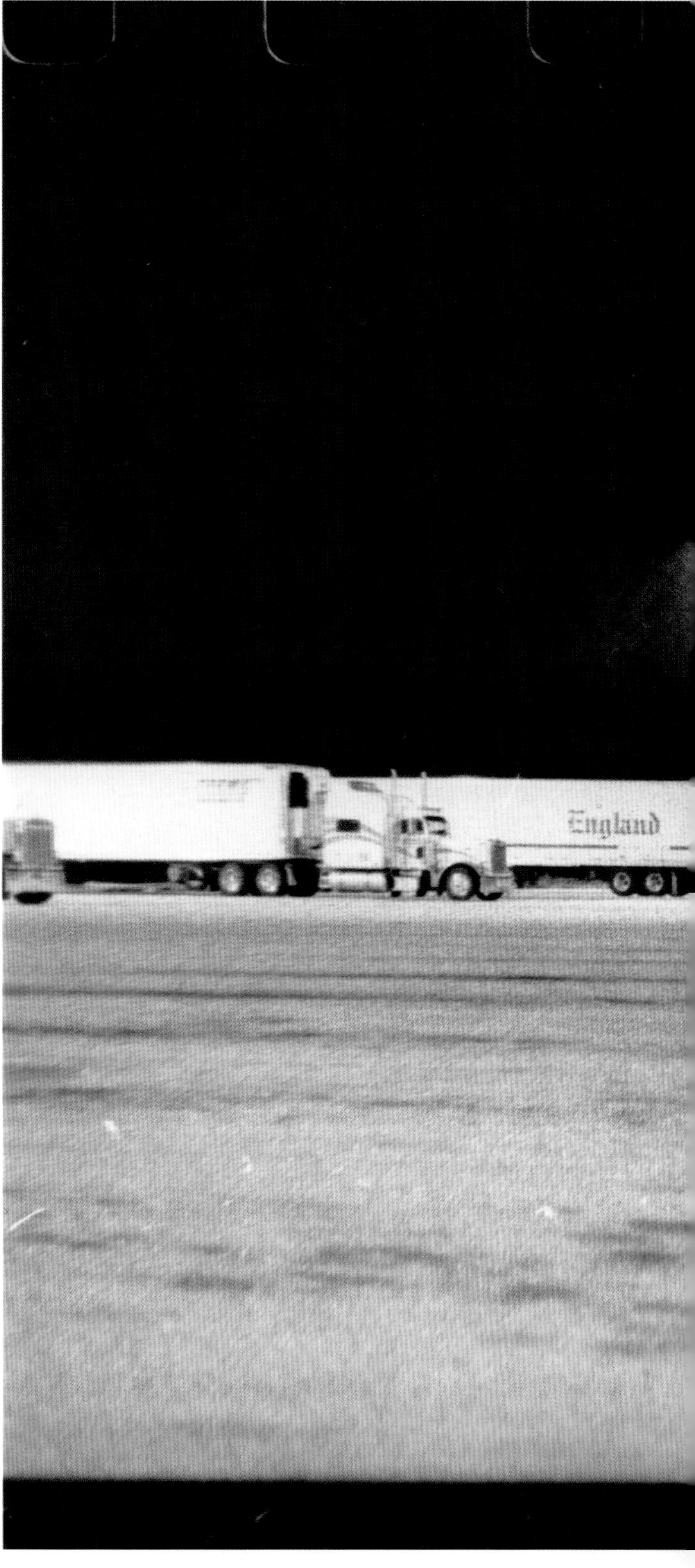

Parking lot at night

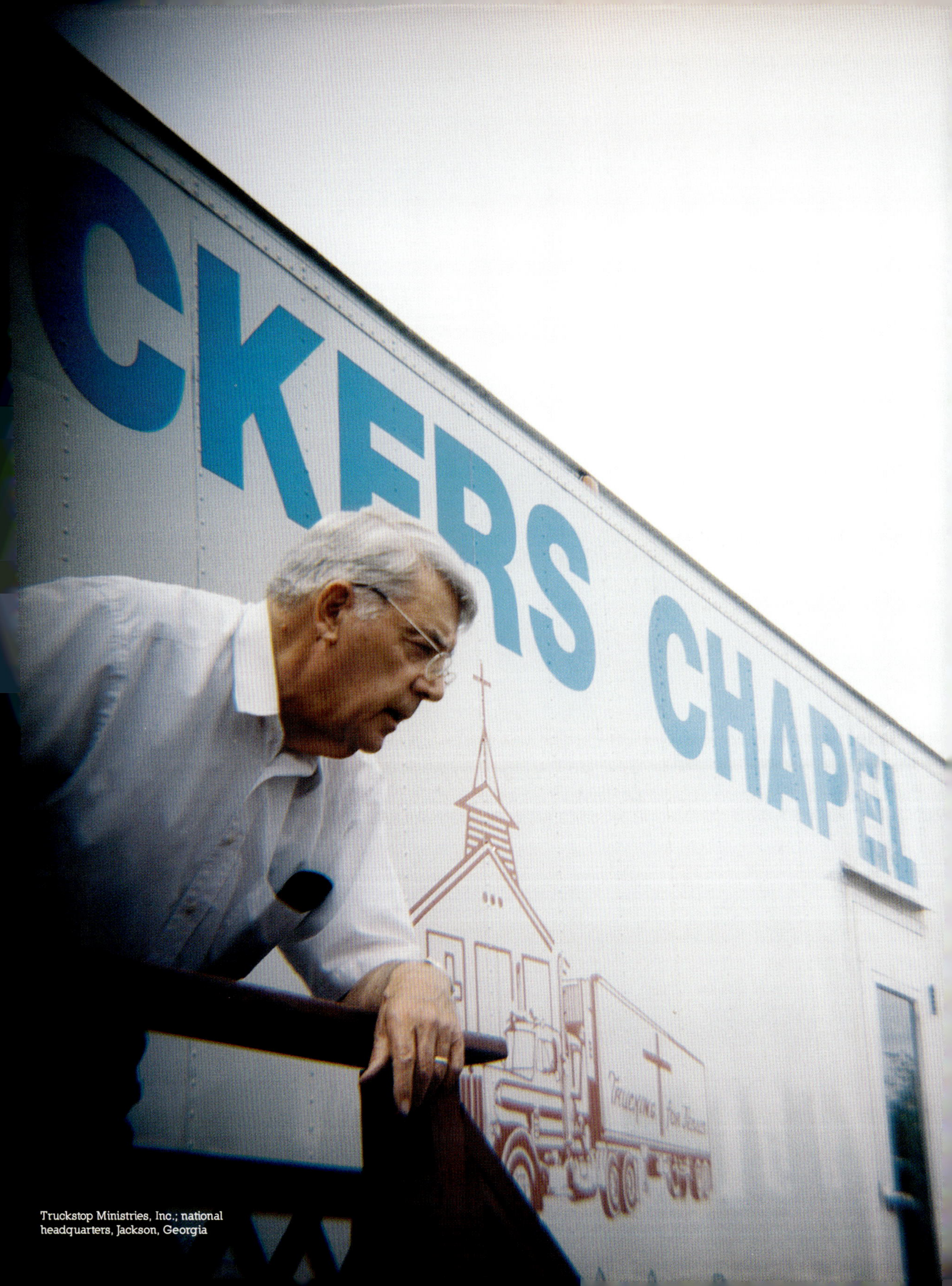

Truckstop Ministries, Inc.; national headquarters, Jackson, Georgia

Here's Hope

Inside the trucker chapel, a prayer book titled *Here's Hope* sits on each chair; they are set in four rows with an aisle between. The Chaplain walks up the aisle and stands behind the podium. He carefully combs his hair, presses stop on the tape player, unscrews the cap and takes a drink from a water bottle, then secures his eyeglasses. "The devil is doing little things to you. Not having enough money for fuel—that's him. He wants to kill you." A knock at the trailer door causes heads to perk up. "Let that man in, it might be someone else that needs money." A bald trucker with a long beard announces that his truck is broken in the parking lot and he needs a ride to a parts store eight miles away. The chaplain instructs him to go inside the truck stop and have the clerk make an announcement over the intercom, then continues delivering his message.

Shontel

Ed sold his truck a year ago because of the cost of fuel. Now he drives a company truck from North Carolina to Missouri and back on a dedicated loop with his puppy, Shontel. He drives; she pants, pees in corners of the truck and naps on his bed. Ed drinks ten cups of coffee a day and consumes three times that in cigarettes. Fifty years of this have left his body tiny and his back slender, hard and hunched in the manner of a thinning piece of driftwood. "When I got out of the navy in '58, that's the only thing that paid. You could earn $150 a week trucking or $50 a week in a plant." Shontel fumbles unsteadily in all directions once she's released from the truck. Ed retired once in 2002. For a year, he stayed at home and tried to enjoy it, but the abuses of the road on a body begin to swell when motion ceases. "I was sick all the time. Might as well work." Shontel pulls on the leash and wraps it around his legs. Ed remains indifferent. Shontel, the warm fuzzy ball of youth, a long ago lover, a tiny package of happiness and affection, packed up with the shaving kit and sliced bread. Shontel, the years that got away from him.

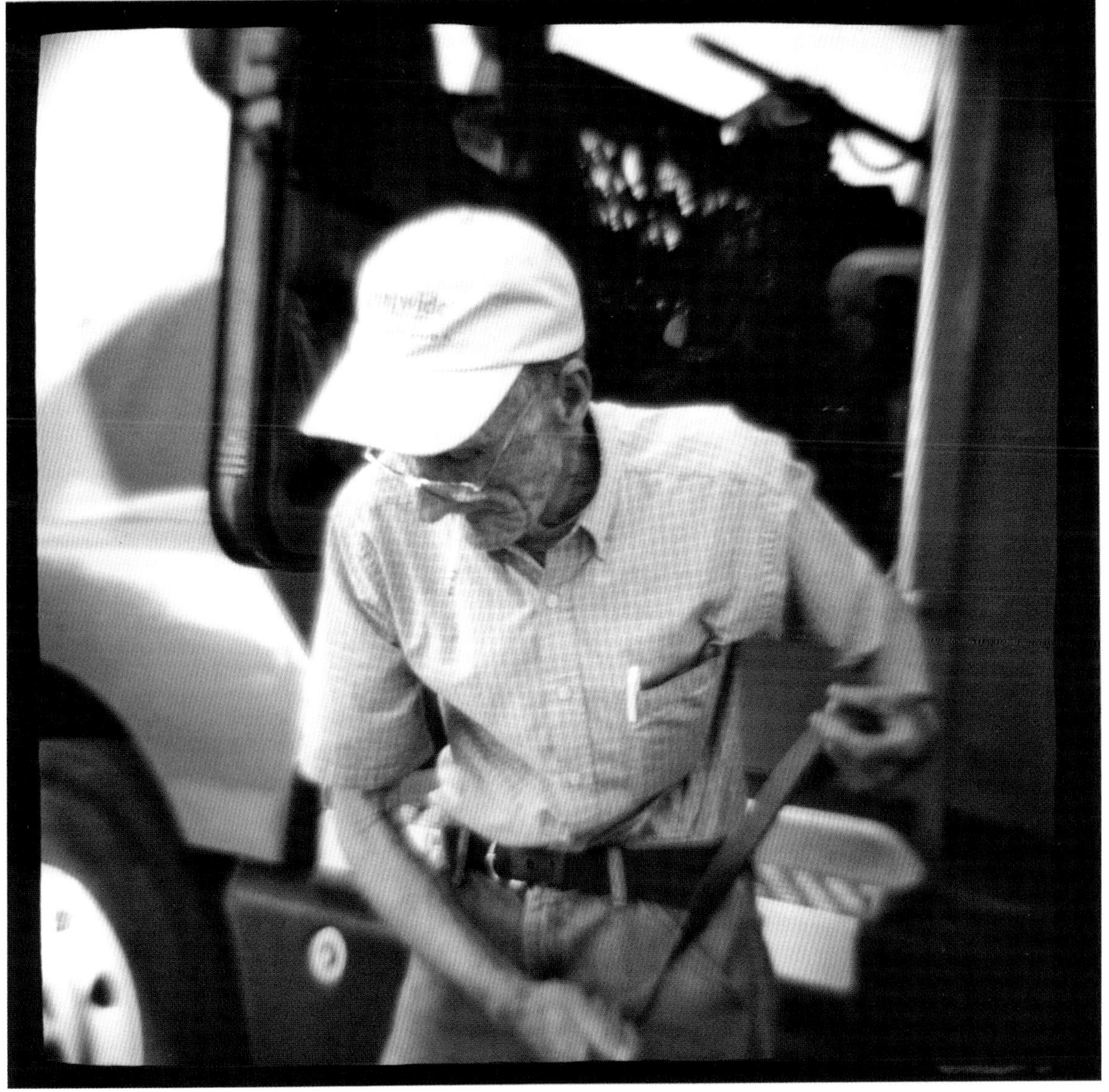

Ed; I-40, Exit 81

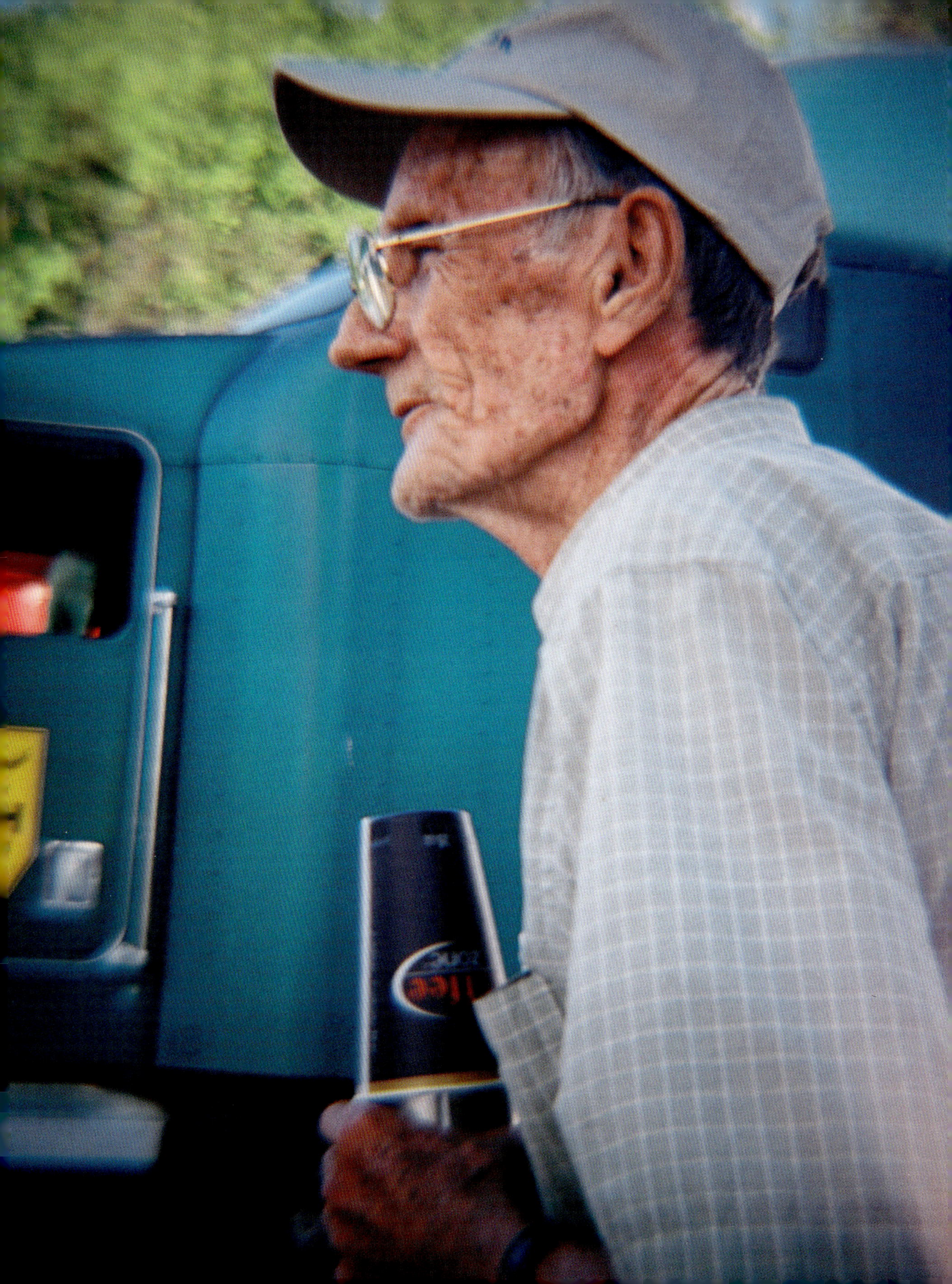

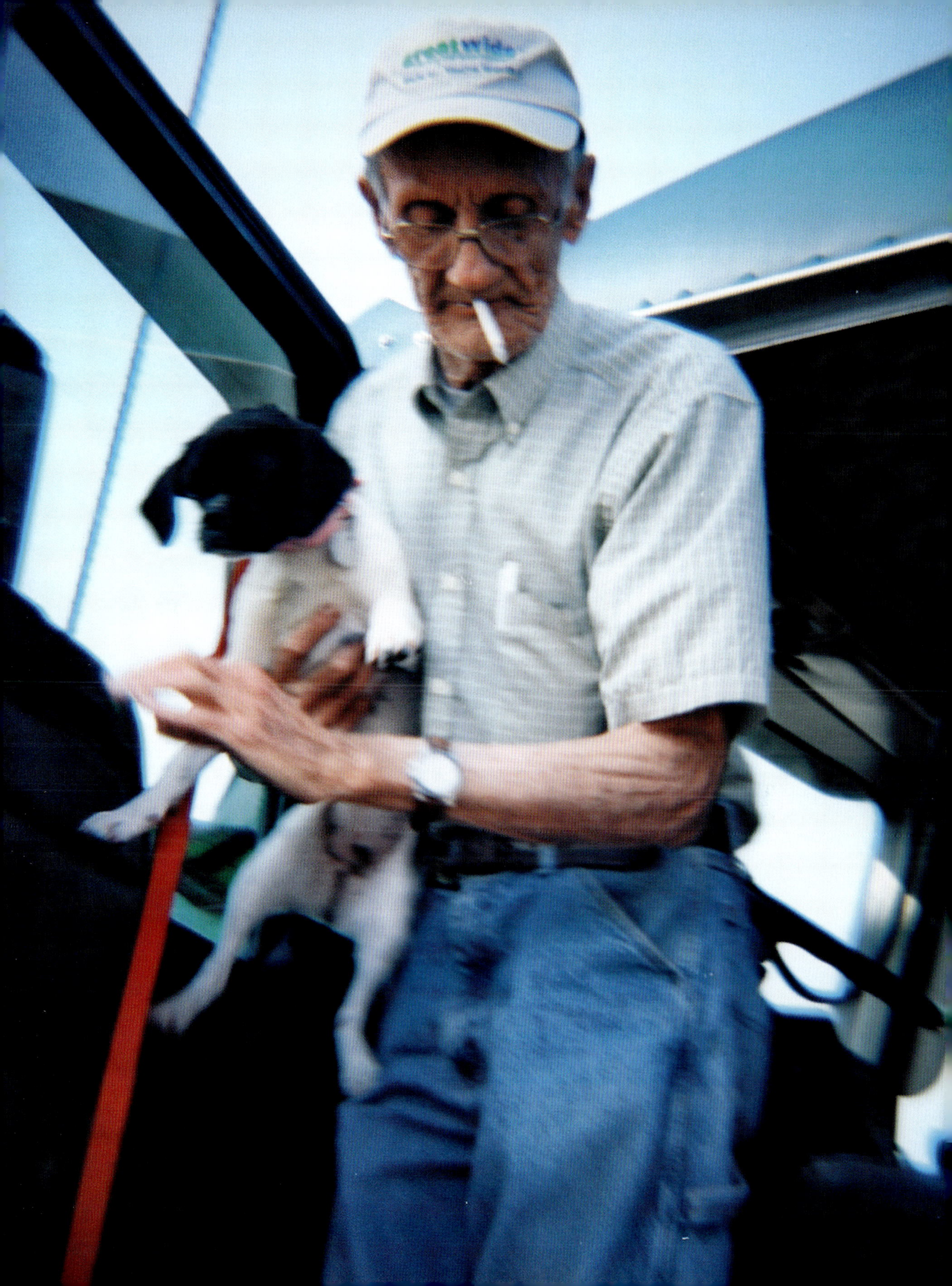

In a Cab, II

Pillow
Remote control
Sleeping bag
Jeans, three pair
Television, flat screen
Boots
Apple juice, one gallon
Trash bag
Coffee mug
Driver's Log
Sliced white bread
Motor Carrier's Atlas
Duffle bag
Cereal
Coffee mug #2
Nyquil
Ibuprofin
Band Aid
Wallet
Instant coffee
Self-extinguishing ashtray
Coffee mug #3
Prescription medication
Flashlight
Highlighter, yellow
Playstation
15 shirts
Easy Cheese, aerosol can
Note pad
Pen, red
Cell phone
Gloves, leather
Calculator

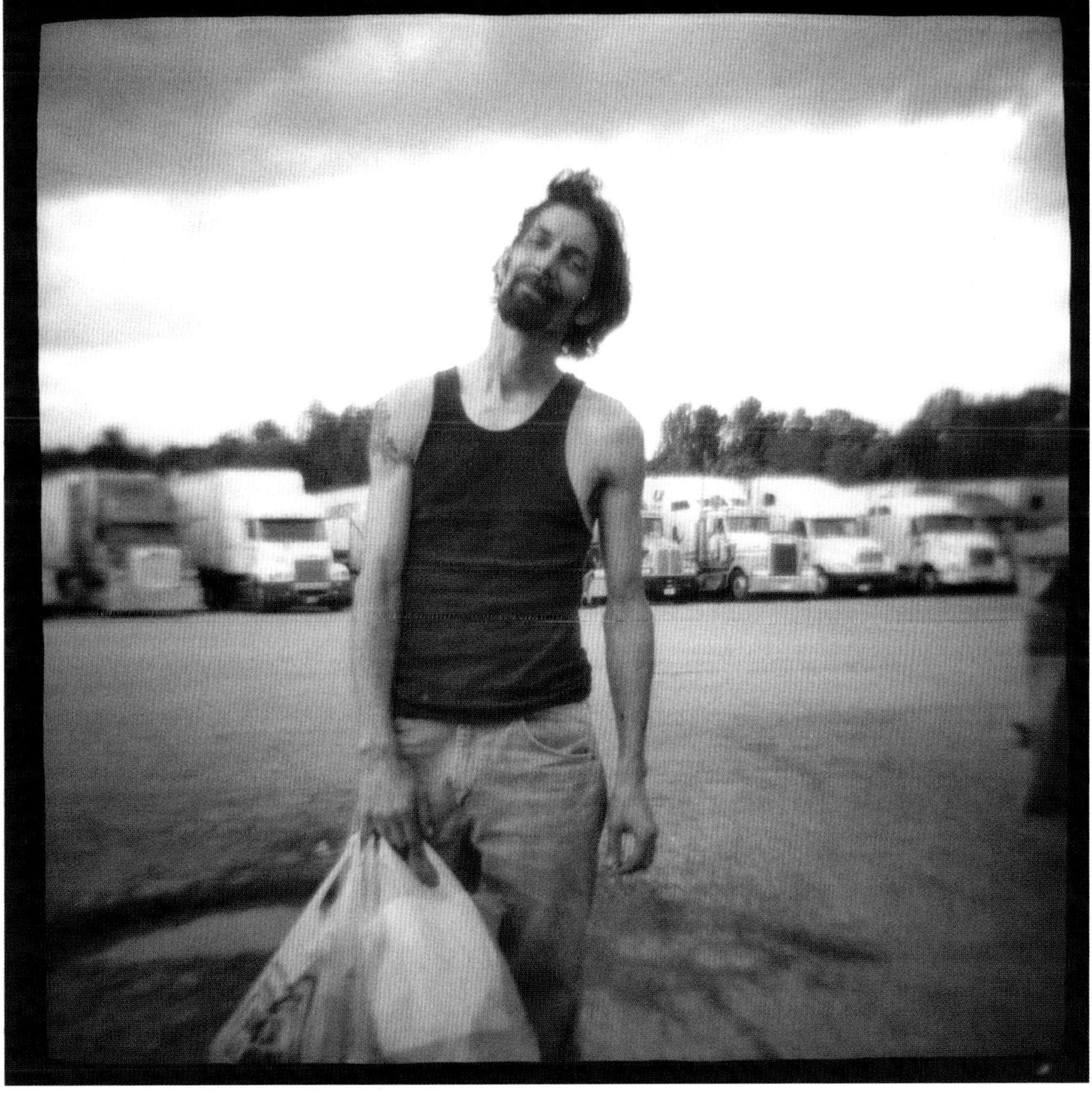

Stopping to do laundry; I-85, Exit 129

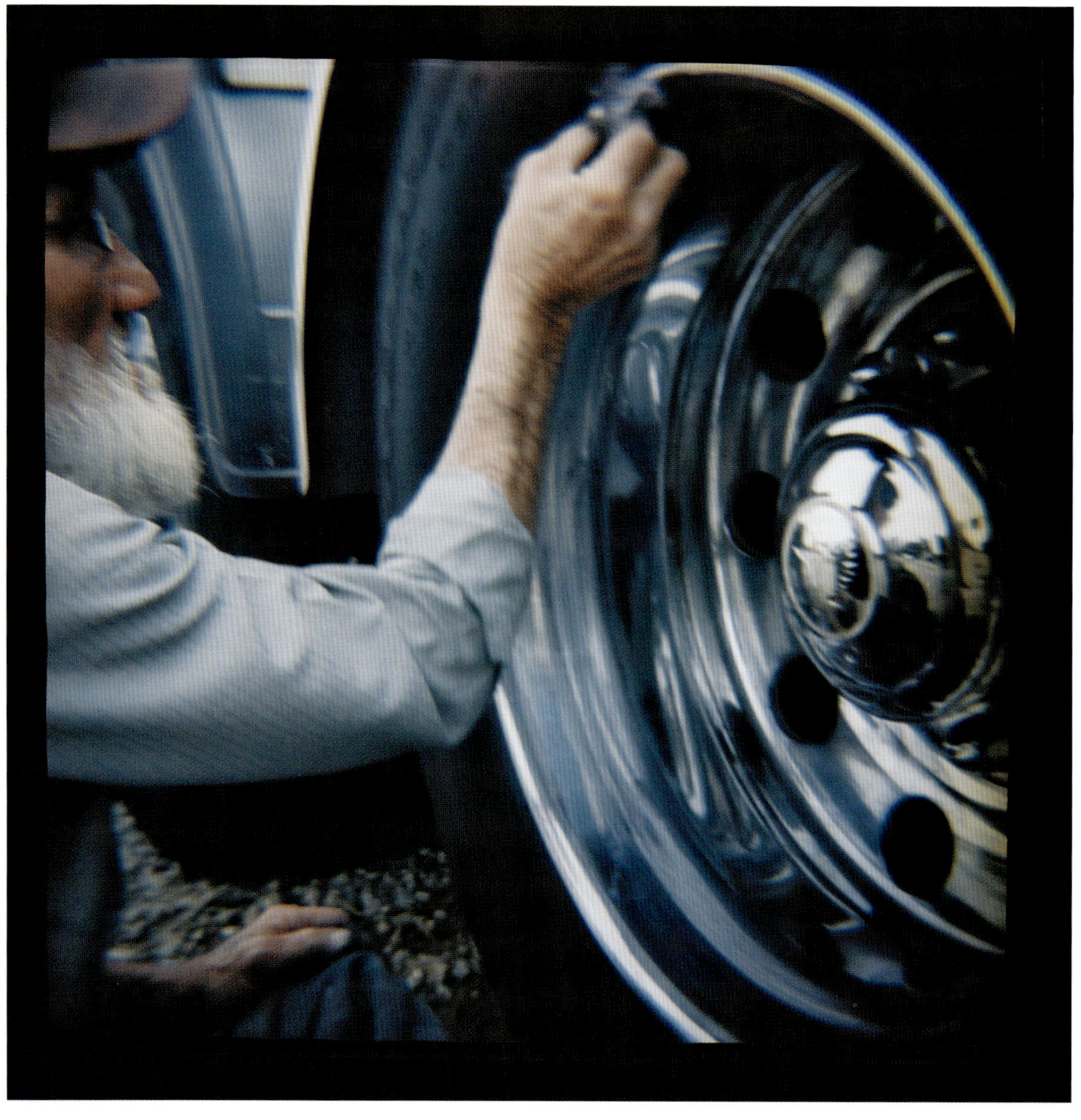

Shining wheels

Kitty Kitty

Kitty Kitty is often mistaken for a prostitute, but she only shines wheels. She lives in a tent behind a massive parking lot where three truck stops converge. It is spring and she will be here until the fall. Her tank top hangs low and her breasts hang out. She is wearing sunglasses and a black baseball cap; her wavy orange-blonde hair crests above the ears. "The homosexual truckers will call the police on me because they think I'm a prostitute and they want all the good buddies for themselves. But the police know me, so when they see it's me, they just keep going." From the furthest ends, the pavement stretches more than the length of two football fields, dipping once in the middle where a second drive allows direct passage to the Truck Wash. The concrete follows the flowing pattern of hills beneath it. It is a lucrative truck stop; at its edges, behind a line of trees, there is also a married couple camping, shining wheels.

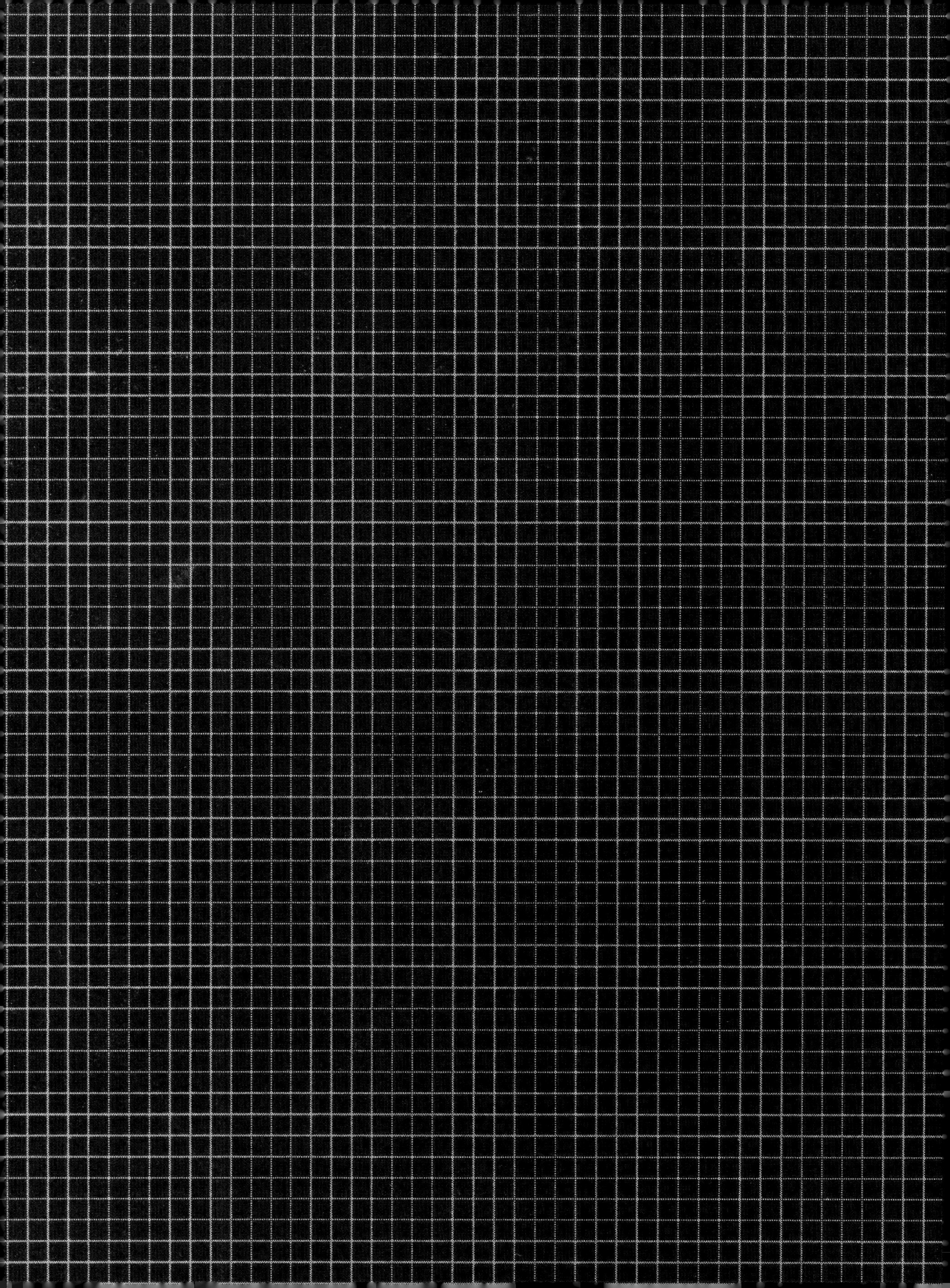

PART THREE
Fuel, costs, labor

In The Morning

Bill Schwartz heard on satellite radio that there has been a 20 percent drop in truckers in the past two months, 90 percent being owner-operators. He gets up, puts on his boots and goes inside to send a fax. Duct tape is stretched over the top of both boots. Blackened and dusty, it matches the leather. He was born in 1973, says he has been trucking for 20 years. "In the state of Idaho, on a farm, the kids can get a temporary CDL for harvest season. At fifteen years old, you can drive an 18-wheeler down the highway, across state lines." He remembers driving about 800 miles a day, waking up at 6 a.m. and driving until 10 p.m., seven days a week. "You get a leave from school for farmer's work." Bill is an owner-operator and leases his truck to a company. He pays for fuel and repairs; they lend him base plates and pay him to drive. "The fuel surcharge for this load coming out here paid me 78 cents a mile on top of the line haul, which was $1.10 a mile. You put the two numbers together and that's what I was making." For two days he has not moved. His truck is parked at the far end of a sandy lot next to a mechanic shop. It cost him $1,600 for a new compressor; the total work will come close to $3,000. An elderly woman with tight curls and orthopedic shoes arrives to pick up his trailer. With a minimal exchange of words, he directs her back and connects the trailer to her truck and manually cranks up the landing gear. She drives away.

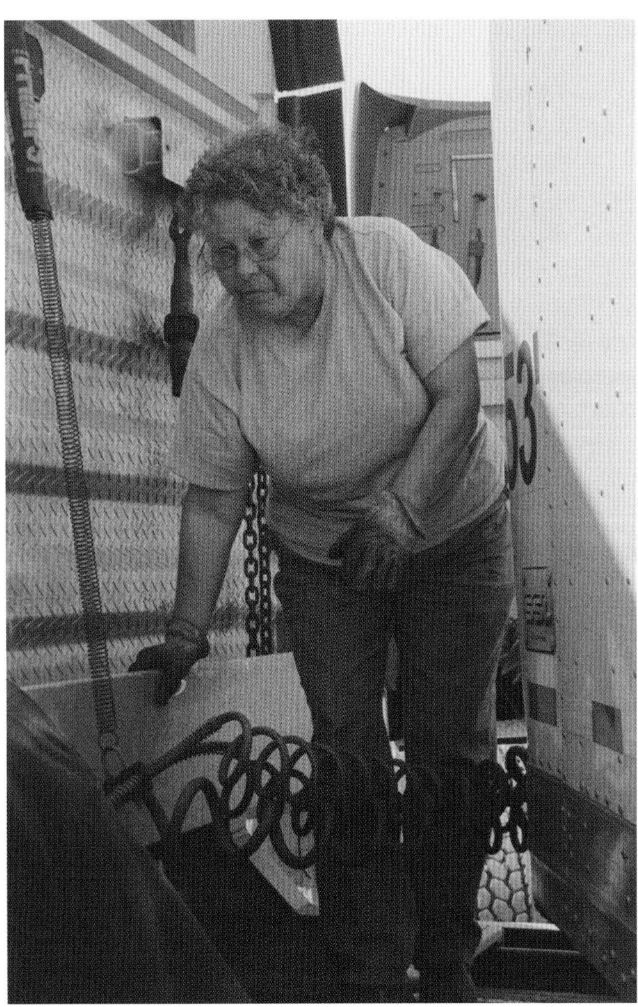

Older woman driver rescuing Bill's trailer

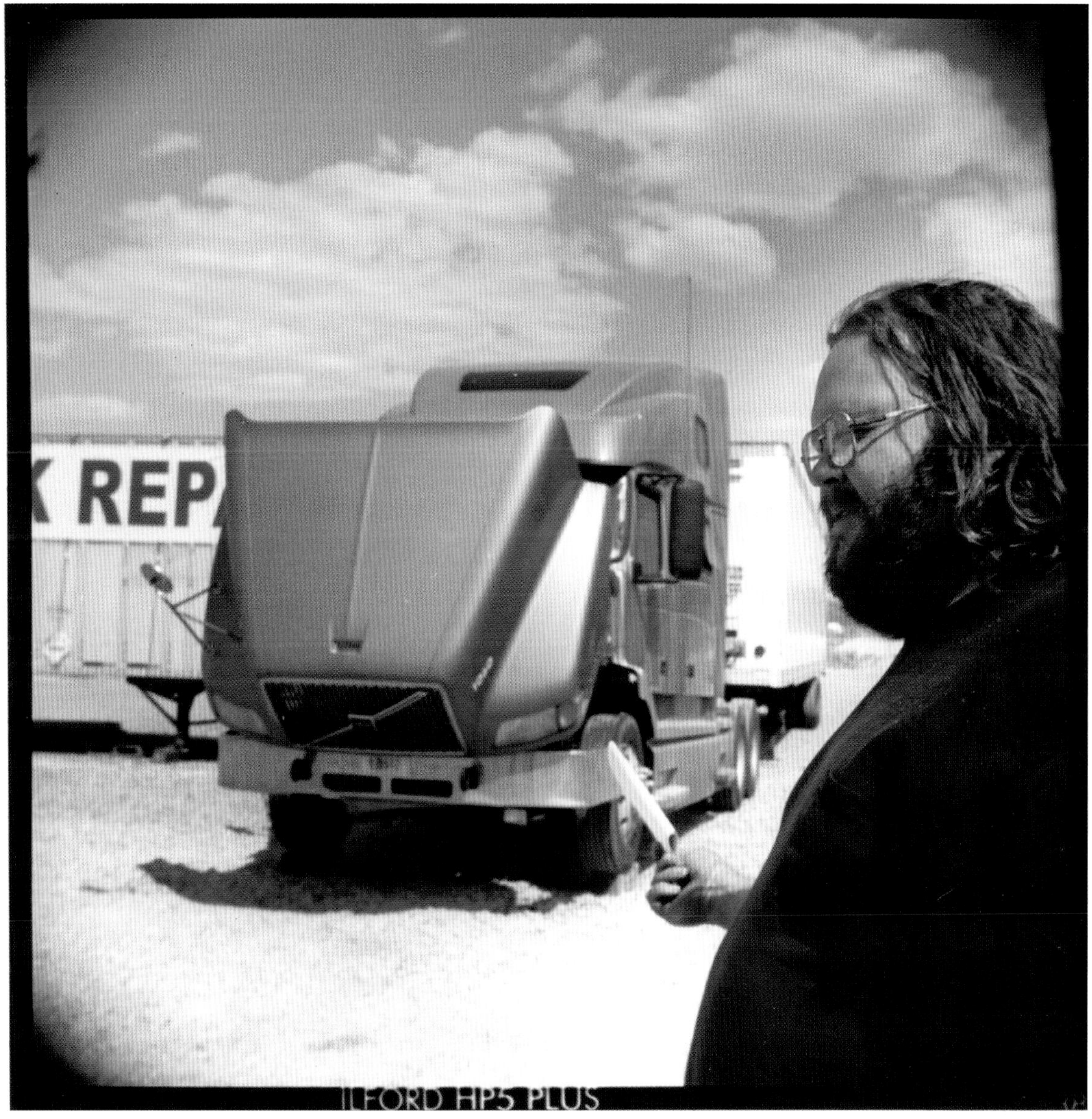

Bill Schwartz; I-80, Exit 352

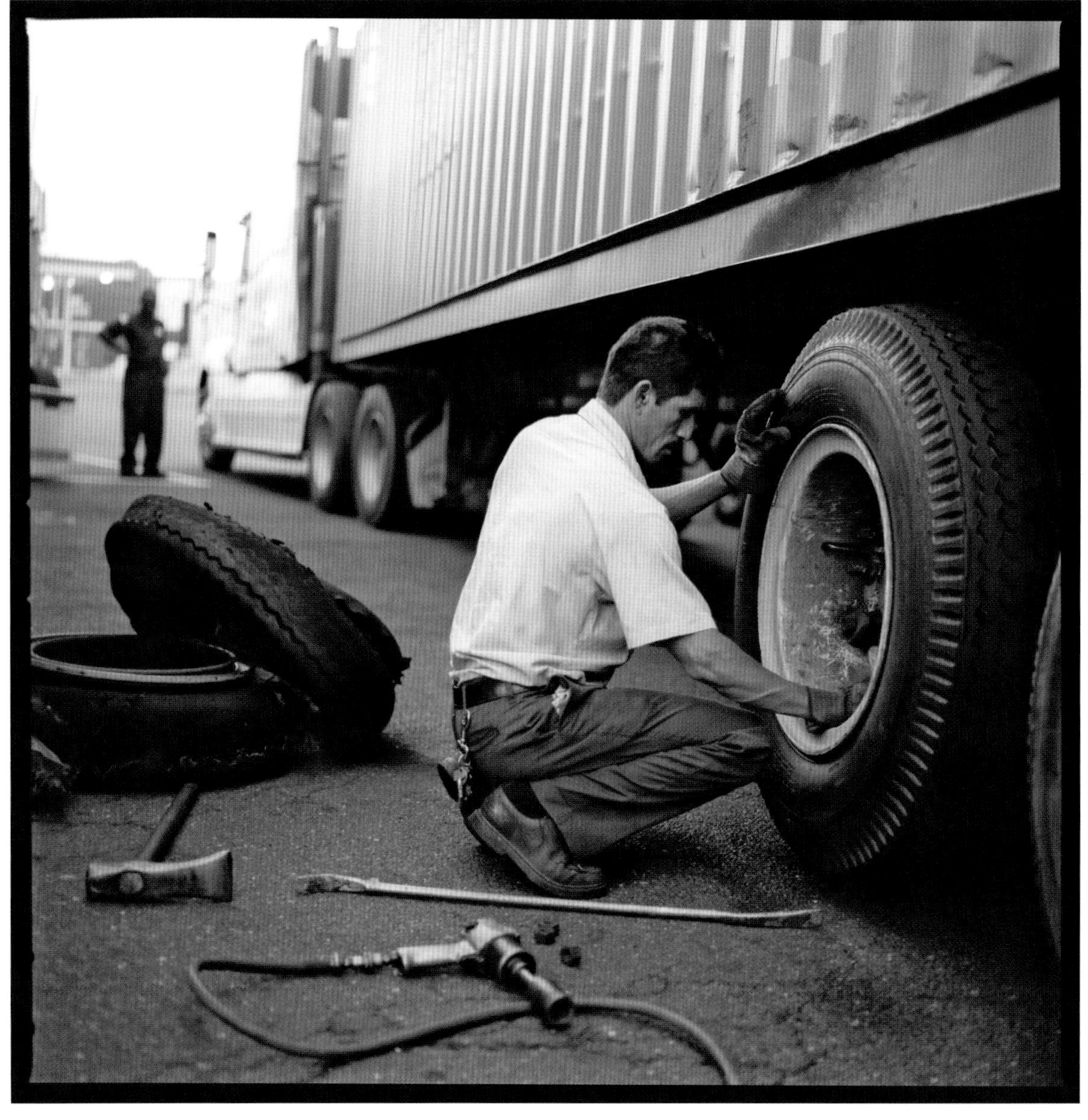

New tire, Modena Service Area; I-87, mile marker 65

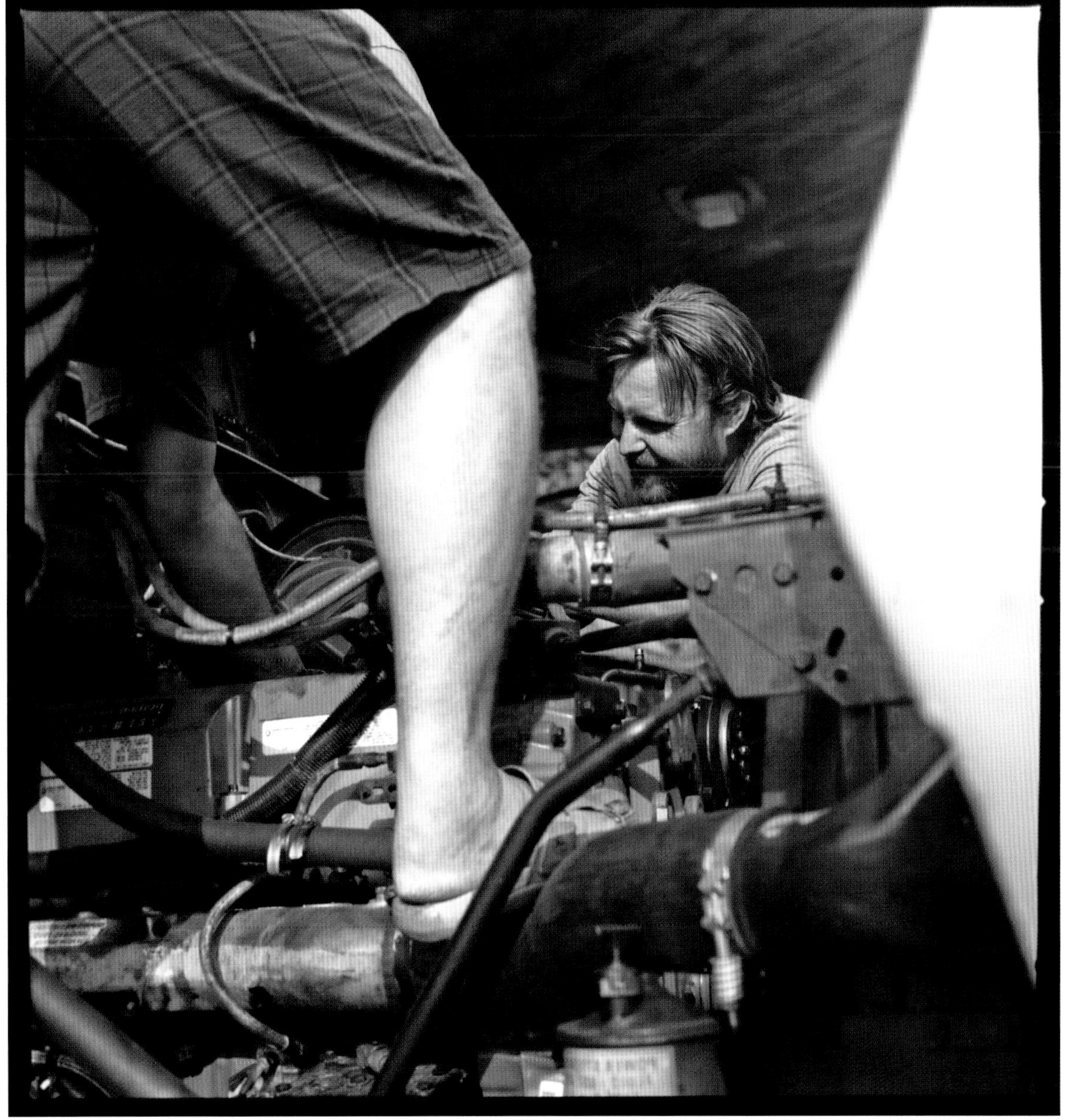

Engine inspection; I-80, Exit 4

FUEL

"There are 240 gallons of fuel in this truck. I fill up three times per week. You have to fill up six or seven times to drive coast to coast. You can drive 800 to 900 miles between fill-ups." Tom Potter's company has 8,000 trucks on the road. At $4.40 per gallon, it costs $1,100 to fill up one tank. On average, trucks get between five and six miles per gallon.

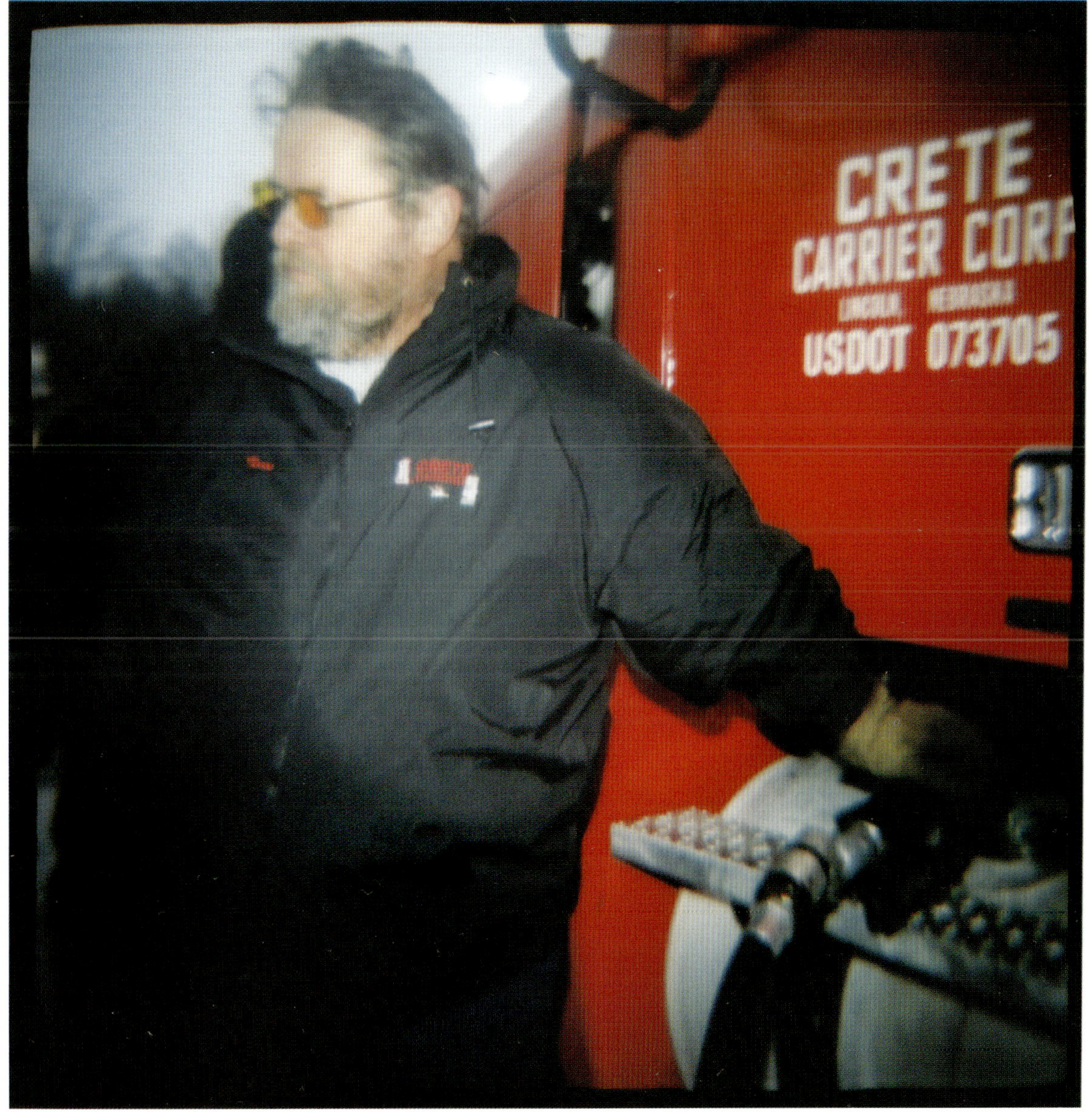

Tom Potter; I-20, Exit 114
b. 1948
Years Driving: 36

Years to Retirement: 11
Average Time on Road: 4-6 weeks,
3 days home

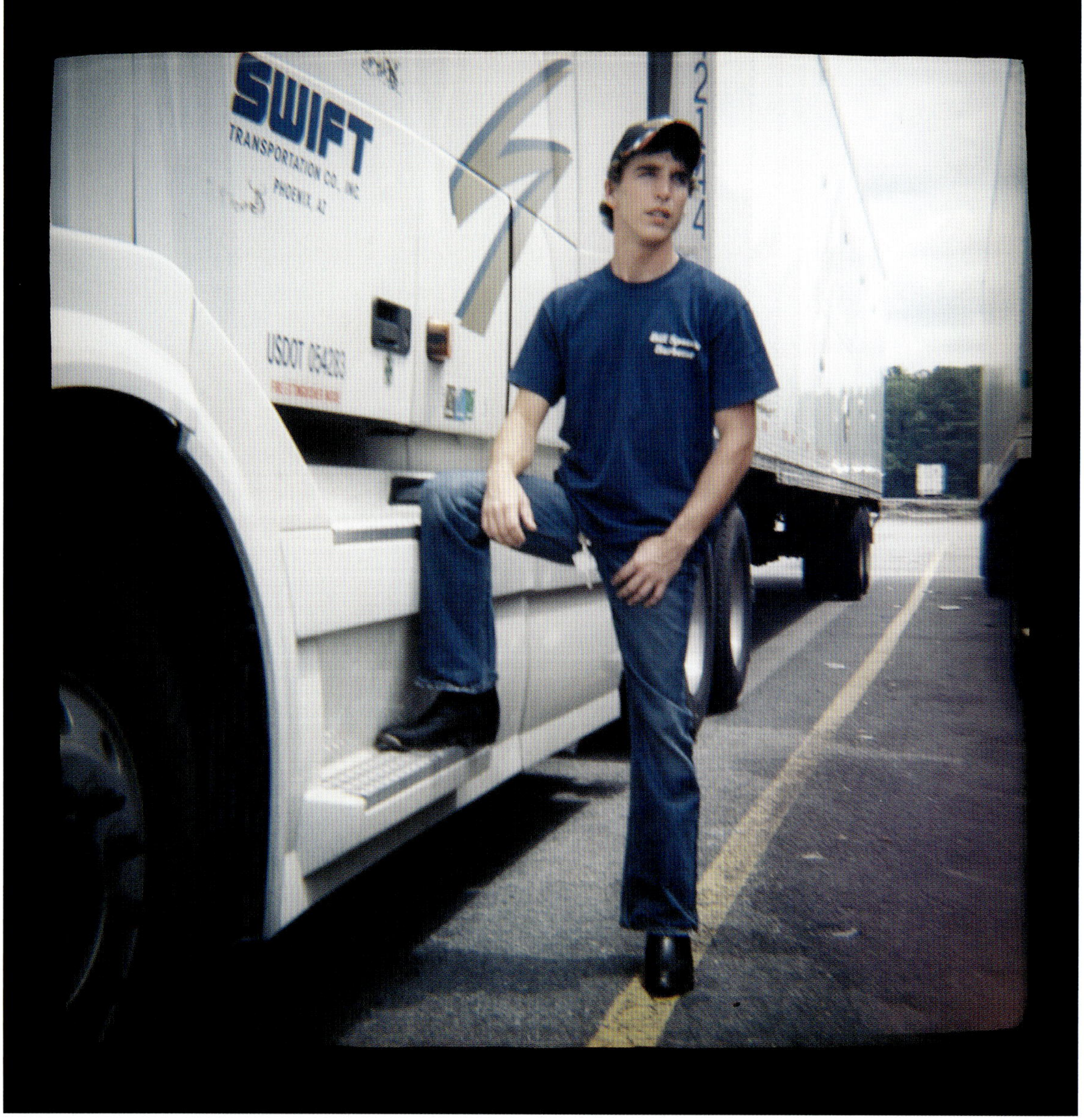

Justin Spoon; I-75, Exit 201

Free As A Bird

"I'm single, as free as a bird." At 22, Justin Spoon weighs 120 pounds. He has been operating heavy machinery in Rockhill, South Carolina, since he was 10 years old. He was hired as a trucker before he earned a CDL. During training, he lost his apartment because $730 for six weeks was not enough to keep it. "It didn't make sense to keep an apartment anyway because I'm only home for a couple days." He spent his savings on tuition and his new employer assisted in financing the other half, about $2,000. The company now pays a large part of his loan payment every month and the difference is taken out of his check. "Last month, they took $250 out of my check and I didn't have any money left. When I called them, they told me they missed a payment the month before so they had to pay two notes." His trailer has to be washed after every delivery because it could have been meat, which might leave behind bacteria. At a truck wash, this costs $24. It comes out of his pocket and he is later reimbursed by the company. It is the same when he has to pay lumpers to unload his trailer at a warehouse. He requests approval of an electronic check through an onboard computer. It comes out of his pay and he is later reimbursed "They always approve it even if I don't have the money." If he has to help unload the freight, he is paid about $25.

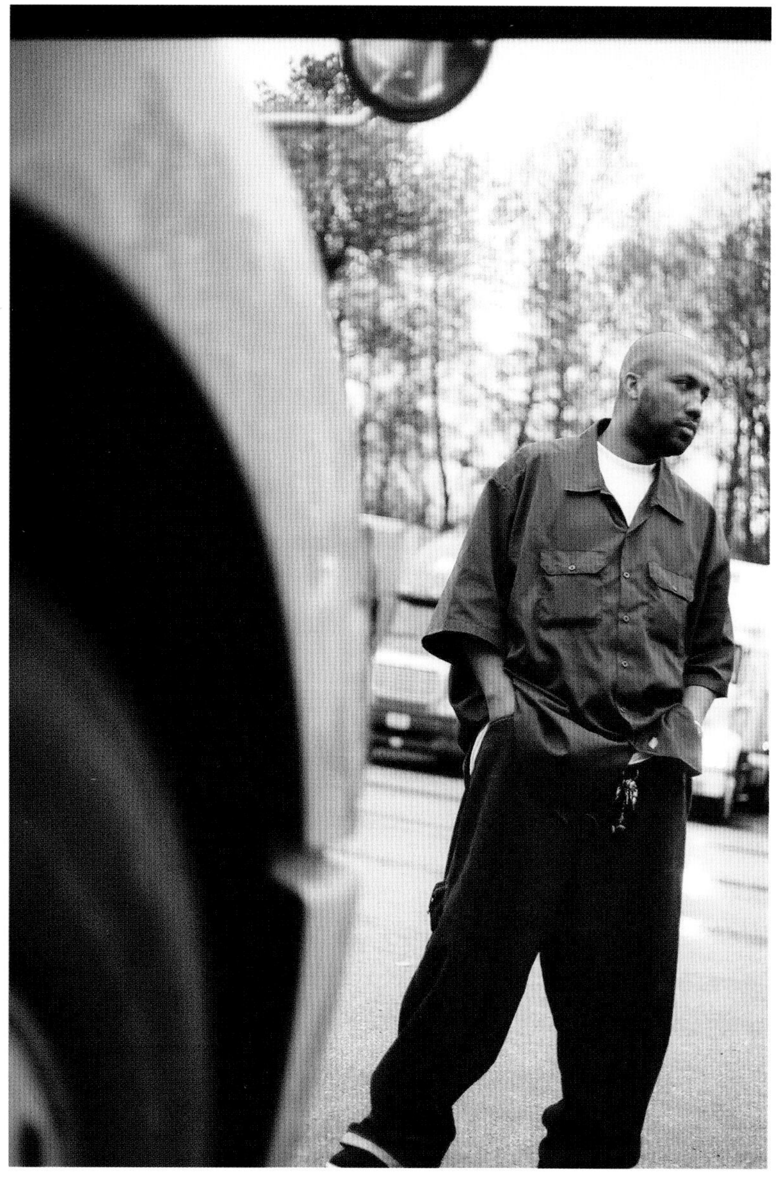

Kevin, company driver; I-285, Exit 51
Years driving: 2
Hometown: Modesto, California

"At the end of the year, I'm going to own my own truck."

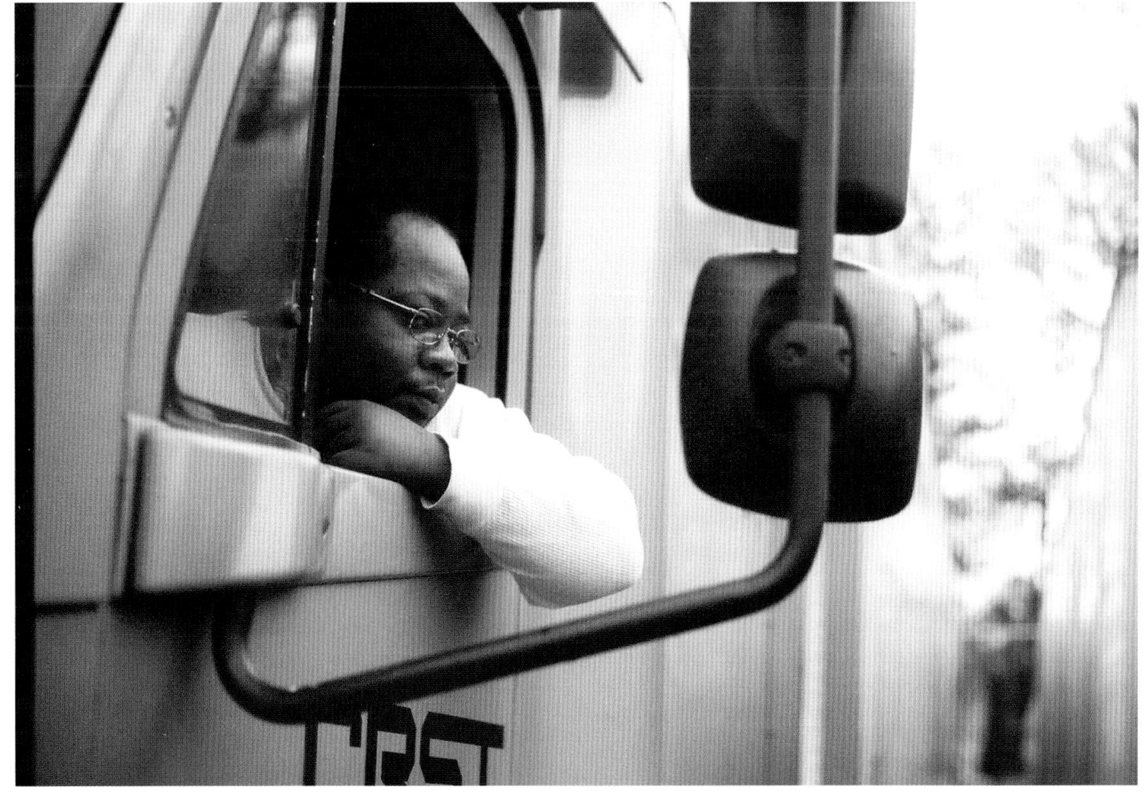

Jeffrey; I-285, Exit 51
Years driving: 8

Family: married, 6 kids
Previous job: restaurant manager

Inheritance

His wide, thick shoulders frame a striped shirt stretched over a hard, round belly. Trucking was his best option as a 16-year-old drop out. His father was a trucker, so he was grandfathered in two years later. "Without an education, truck driving was handed down to me." His wife rides with him but does not split the driving; standing inside the cab, she places a pair of dirty socks over the ledge of the window and rolls it up, then ducks behind the curtain. Their family is grown so they ride together during the week and go home on weekends. He refers to this dedicated route as "a gift from God."

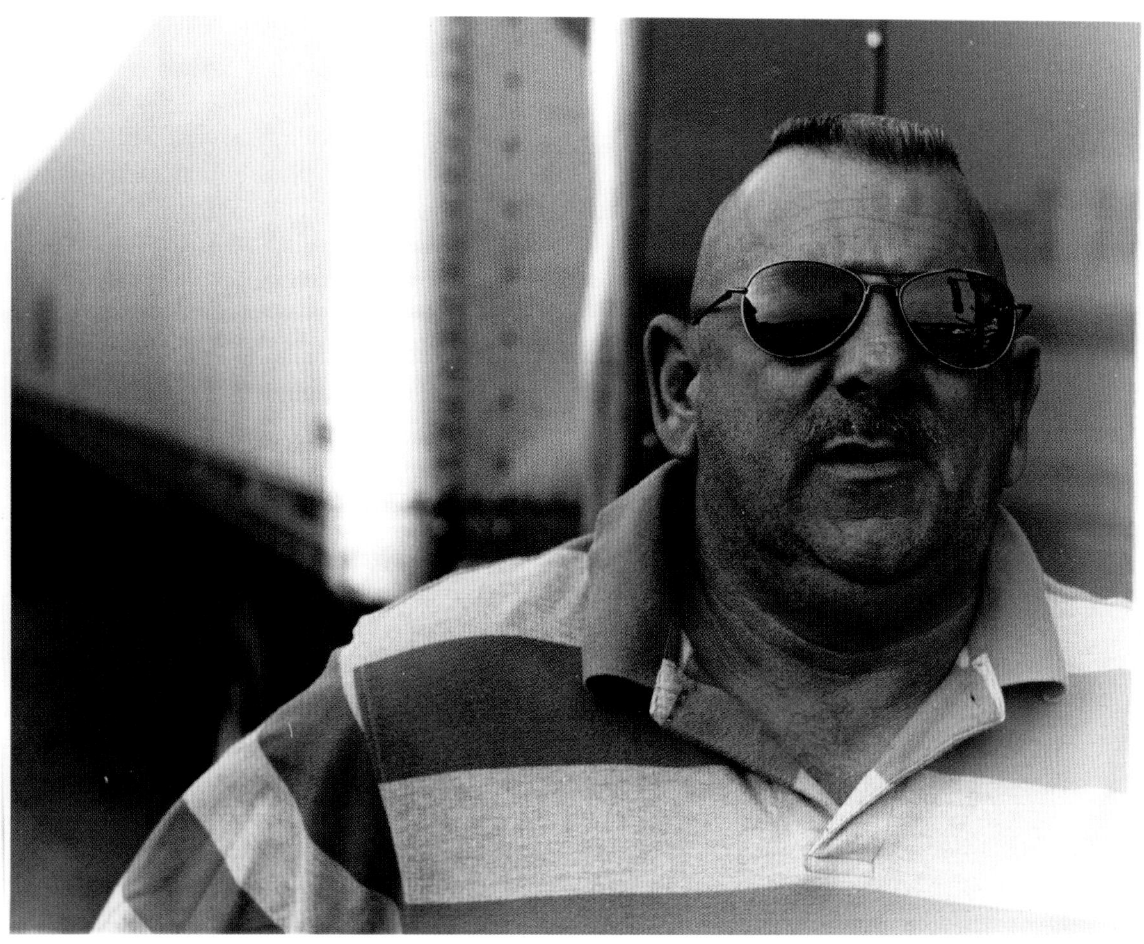

Marvin Turner; I-85, Exit 42

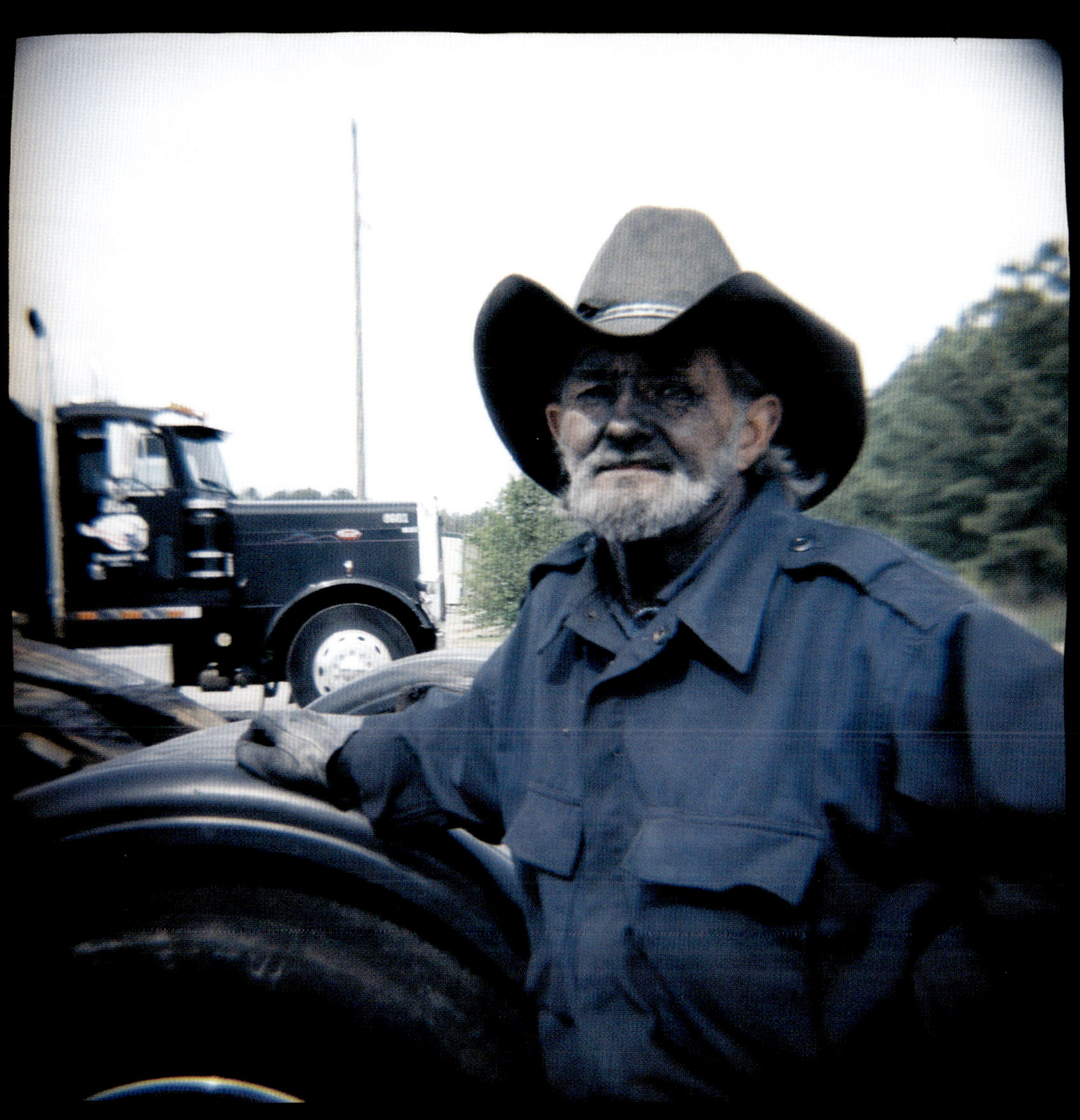

"They call me Cowboy"

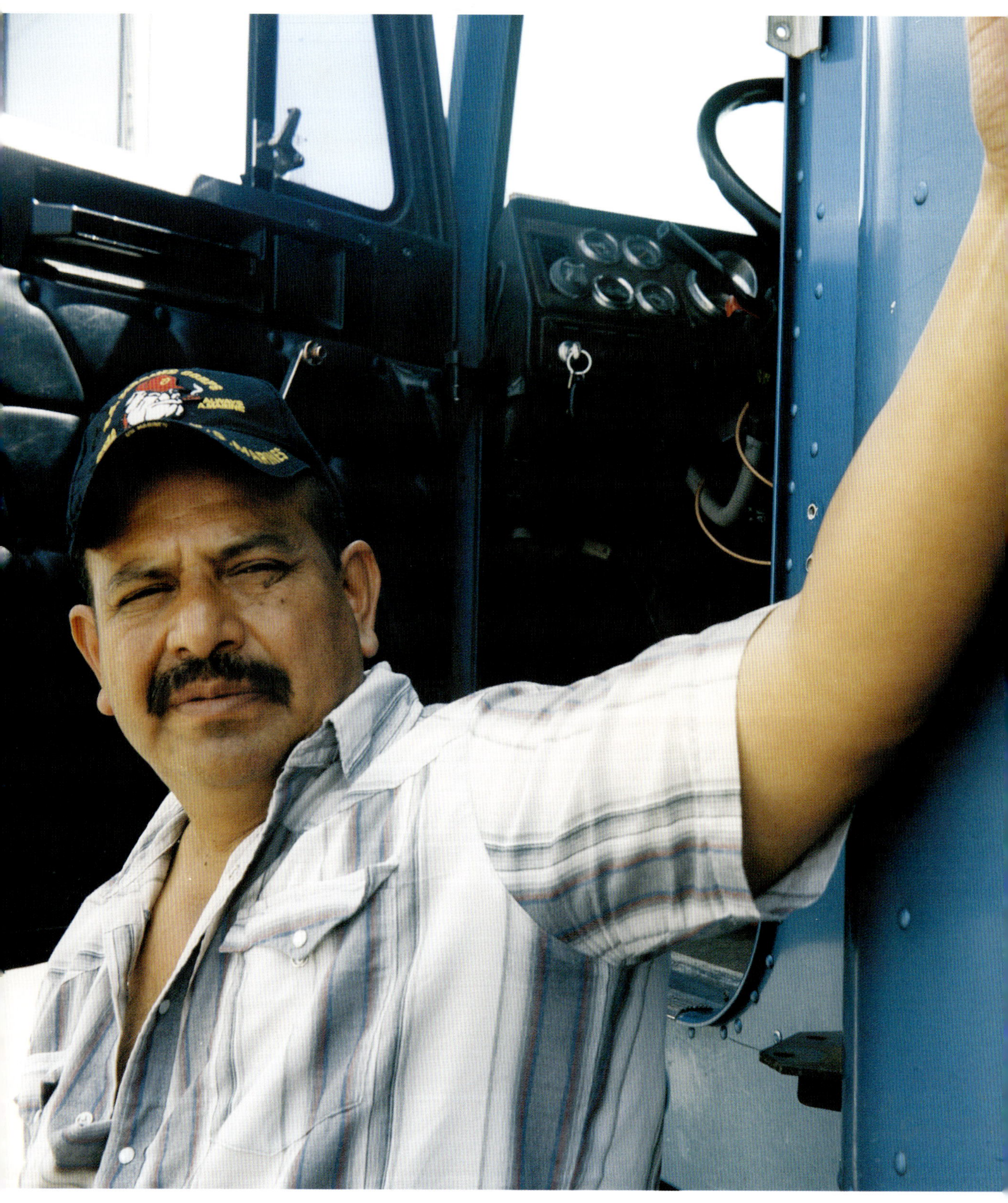

Saul Saldaña, owner-operator, taking a break; I-80, Exit 229

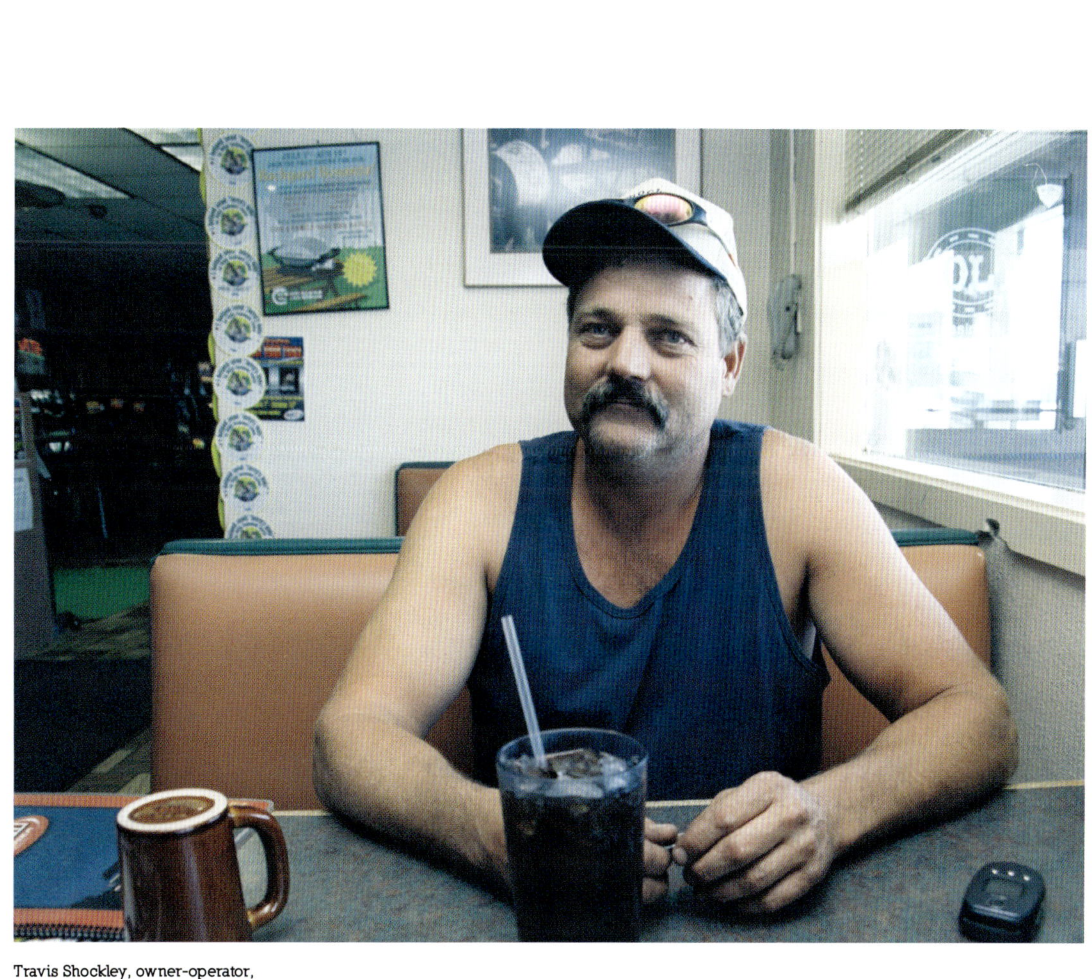

Travis Shockley, owner-operator,
ordering chicken-fried steak;
I-80, Exit 229

Letting Go

"We sold our house and put our stuff in storage. This is the first time we've ever been able to save money. No kids, no rent, no utilities. Room and board are paid for. We went home at the end of April. That was the first time we'd been home since Christmas." Mark had spent the previous 19 years in a spring factory, making springs for air compressors and freezers. They deliver everything from televisions to alcohol, *French's* mustard to *Frank's* hot sauce.

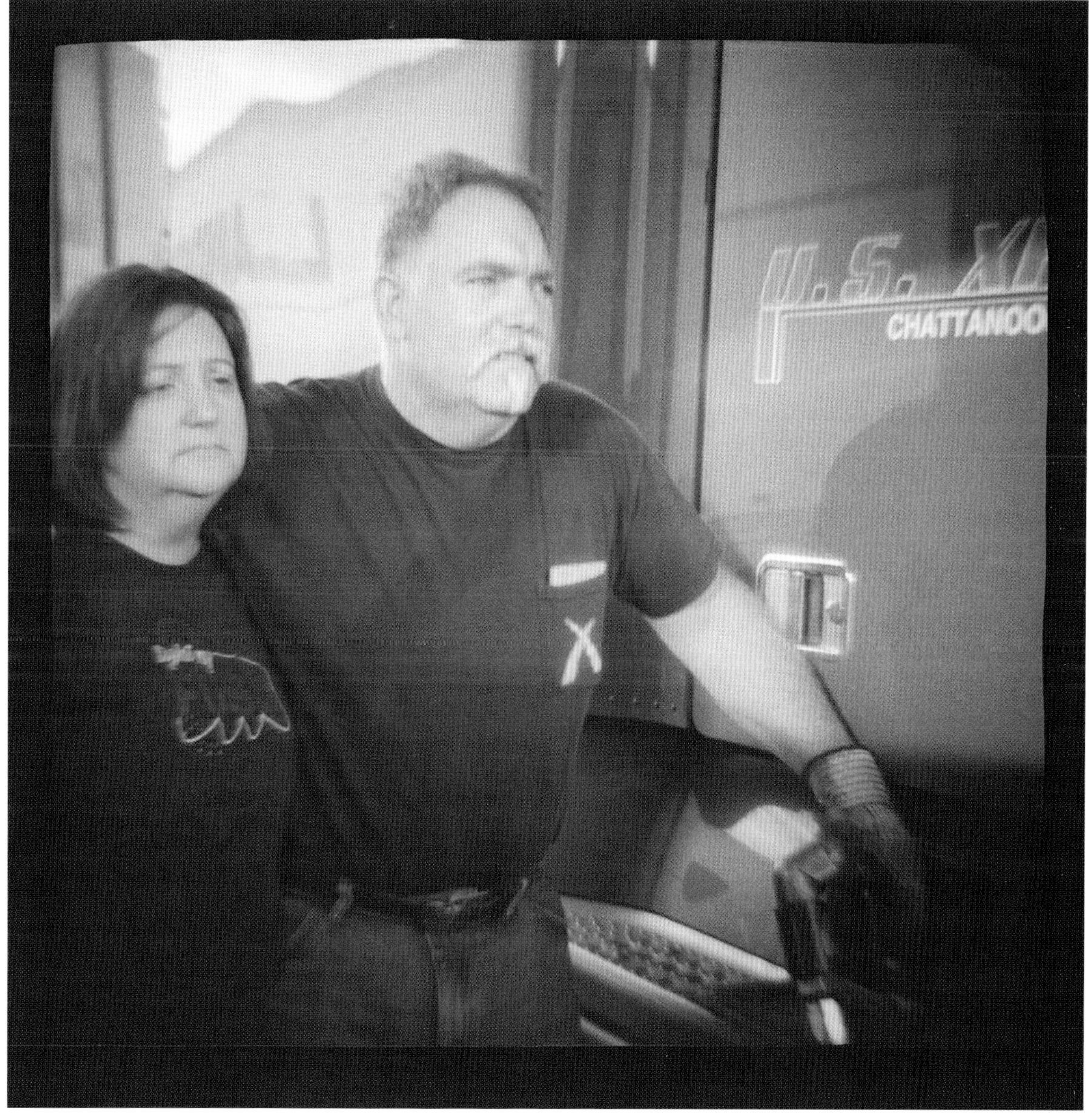

Ellie and Mark Tope; I-85, Exit 129
Years driving: 3

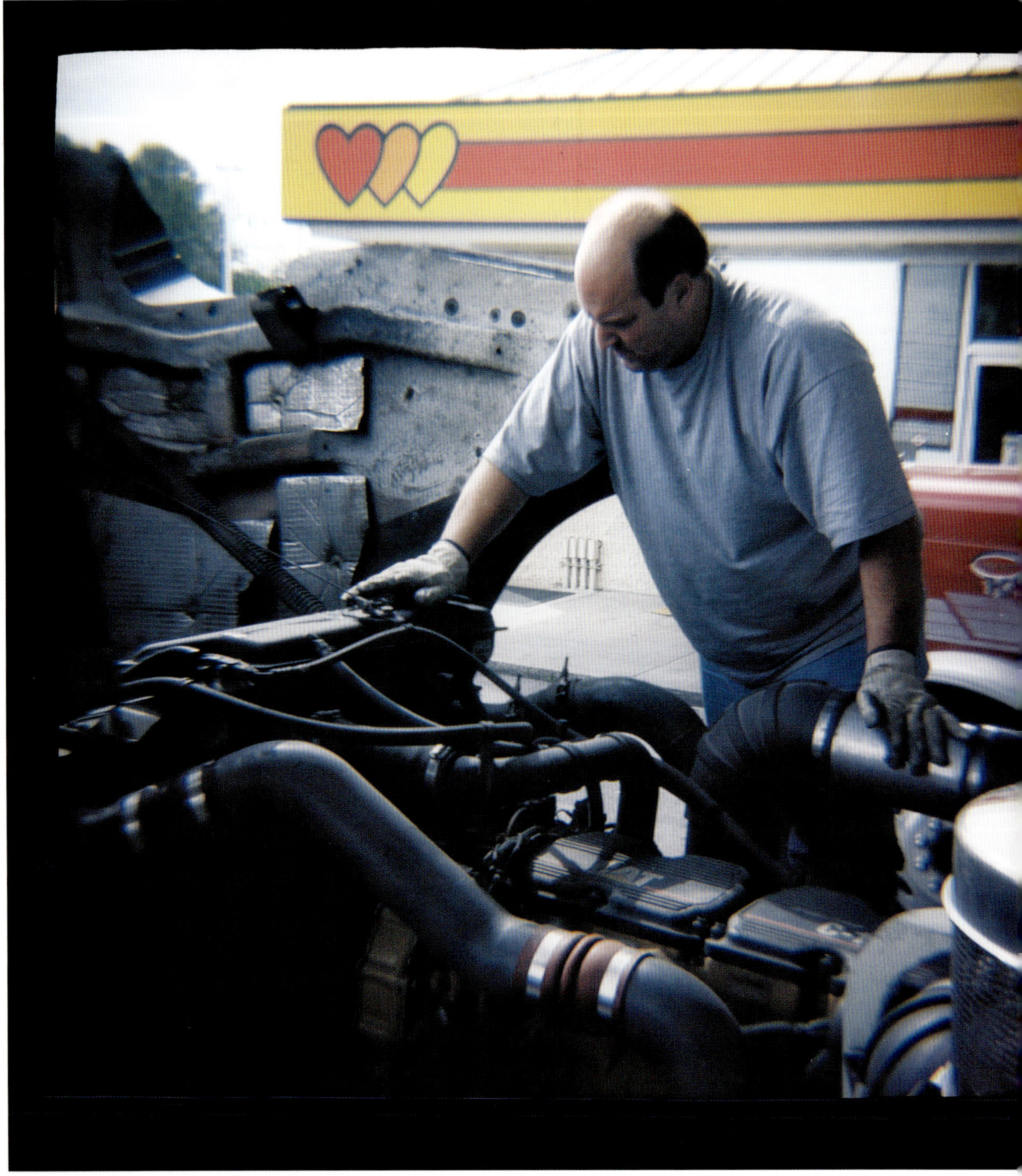

Checking the engine; I-75, Exit 201

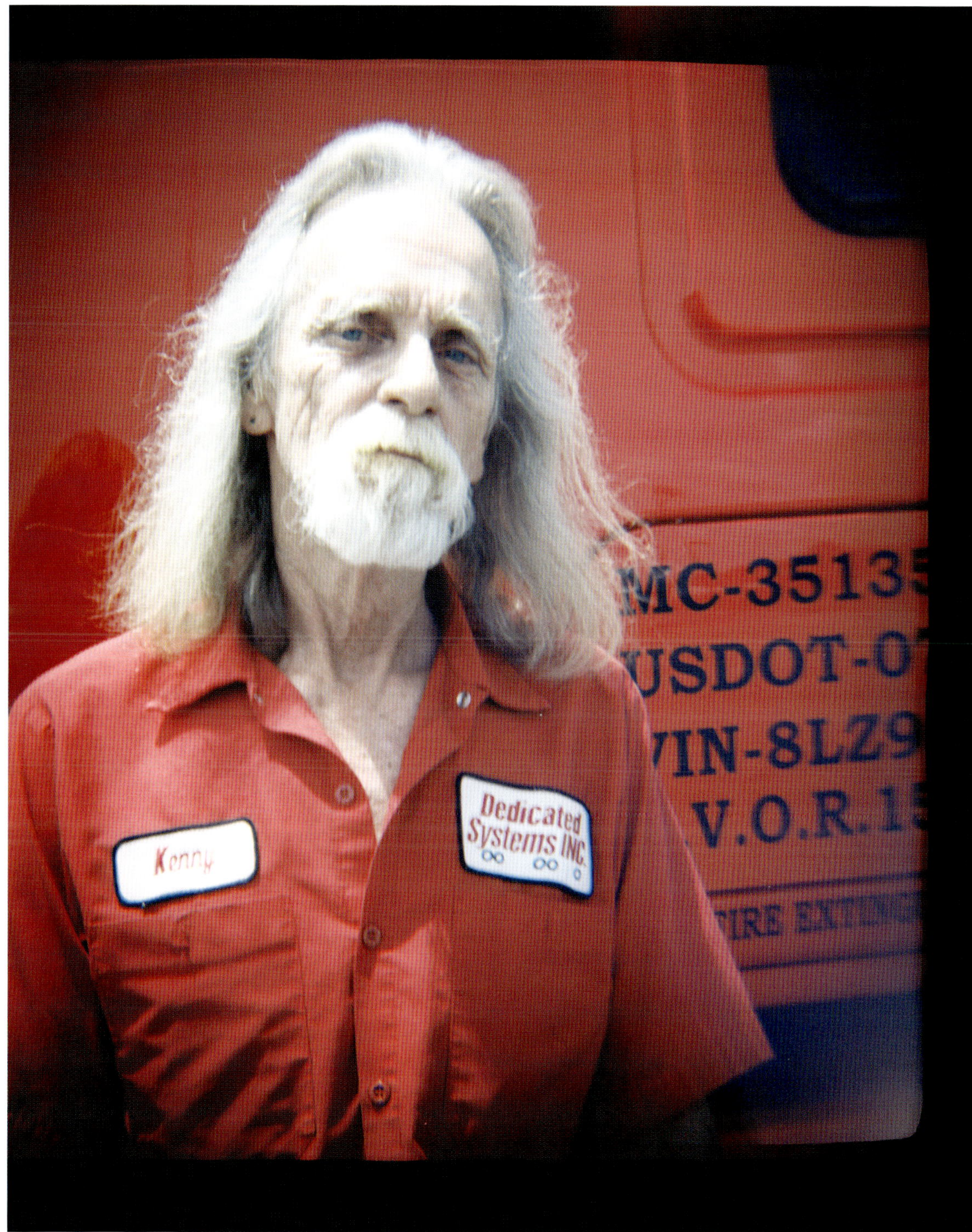

Kenny Davis, delivering chairs for a school

A Mess Of Things

Kurt is delivering "a mess of things from General Mills to Wal-Mart." He says one company is not allowed to monopolize the cargo shipped by Wal-Mart; it has to be distributed between several trucking companies. Wal-Mart has its own fleet of trucks and hires its own drivers to bring merchandise from its distribution centers to stores.

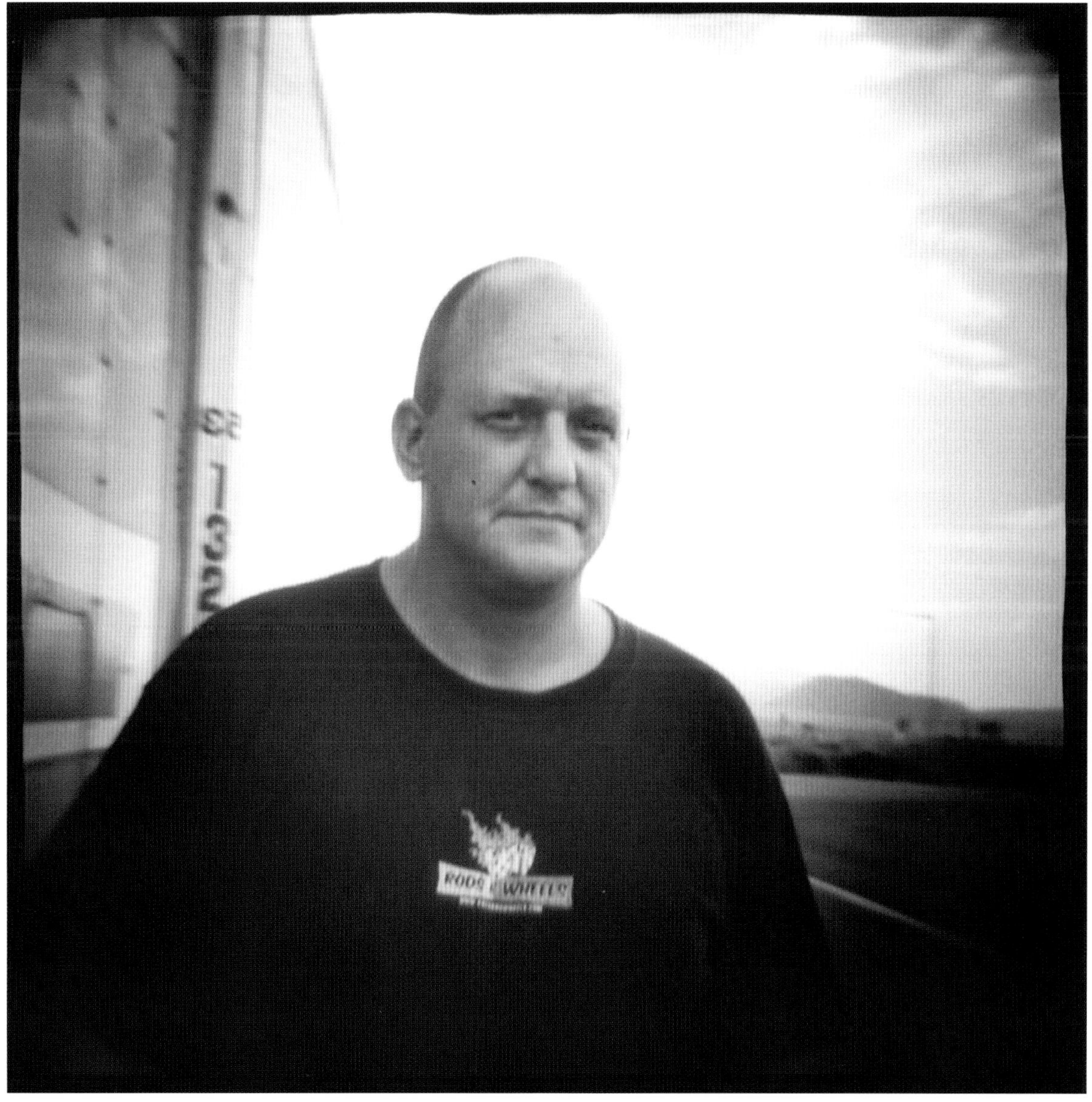

Kurt at a rest area; Hwy. 295 near
Redding, California

Bubblin' Crude

Clark Goodwin, a trucker of 27 years, is parked at a truck stop casino playing *Beverly Hillbillies Bubblin' Crude*. He hauls crude oil used for black top.

"If you want to try to save your truck, you're going to do what you got to do. If people weren't out there cutting each other's throats…if everybody says 'No, I'm not going to haul it for less than three bucks a mile,' then they'd have to pay it. But you've always got some ding-dong out there, 10-thousand trucks strong…they can haul it for a buck a mile. It hurts everybody."

Is this a regular stop for you?

"No, I run all over the place, wherever they're doing black top."

Is black top made with crude?

"It's trash of the oil they make the black top out of."

Is there a big demand for it?

"Oh yeah, they've got to build the roads, keep the roads up. The demand's there but cement and black top's all the same price. Black top's getting more expensive pretty quick. It's up to $1,400 a ton.

"It's a real mess out there right now…there's no control over the fuel prices, so you do what you have to to stay in business until it's all over. Then they just take you away, they take your whole life. It's more of a dream of everybody that's out there to own your own truck, to one day have your own truck."

Do many companies have their own internal leasing programs?

"When somebody leases on with an option to buy, they run them to death. The only thing they're worried about is if that guy is going to make the payment on that truck. They control the money and before the guy gets paid for what he does, they take their money first. I bought my first truck 27 years ago. I had 10 at one time, now I'm down to one truck. I've got two trucks parked at home right now. They're paid for, but I'm only breaking even. I'm not making money right now. You can't be out here turning wheels to break even."

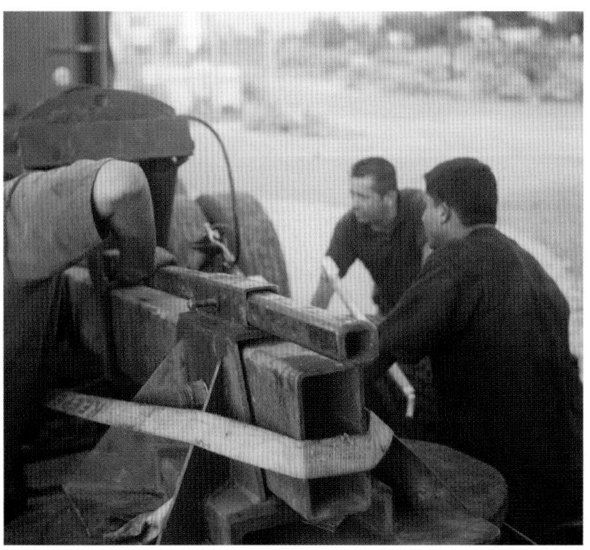

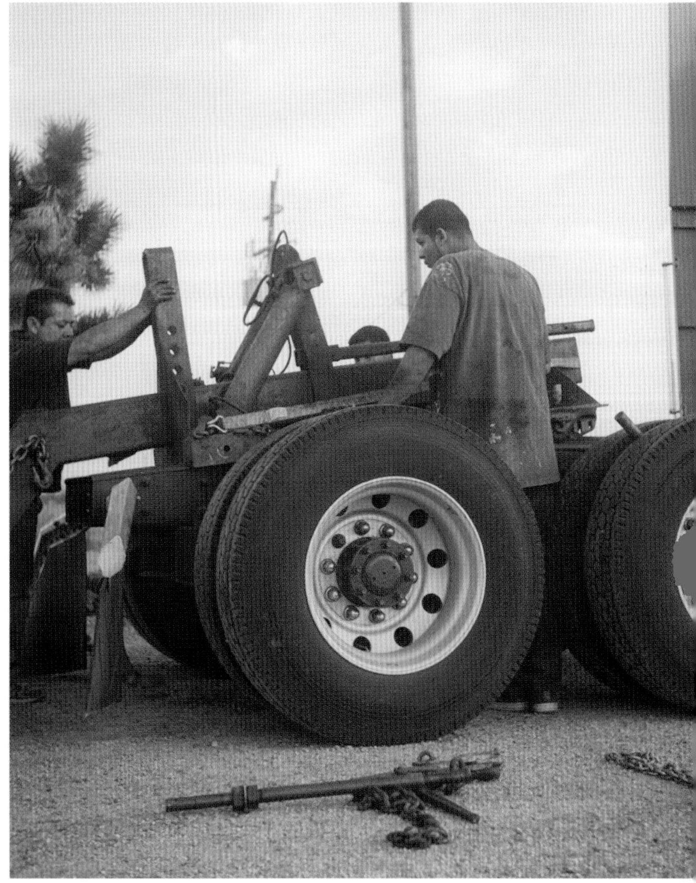

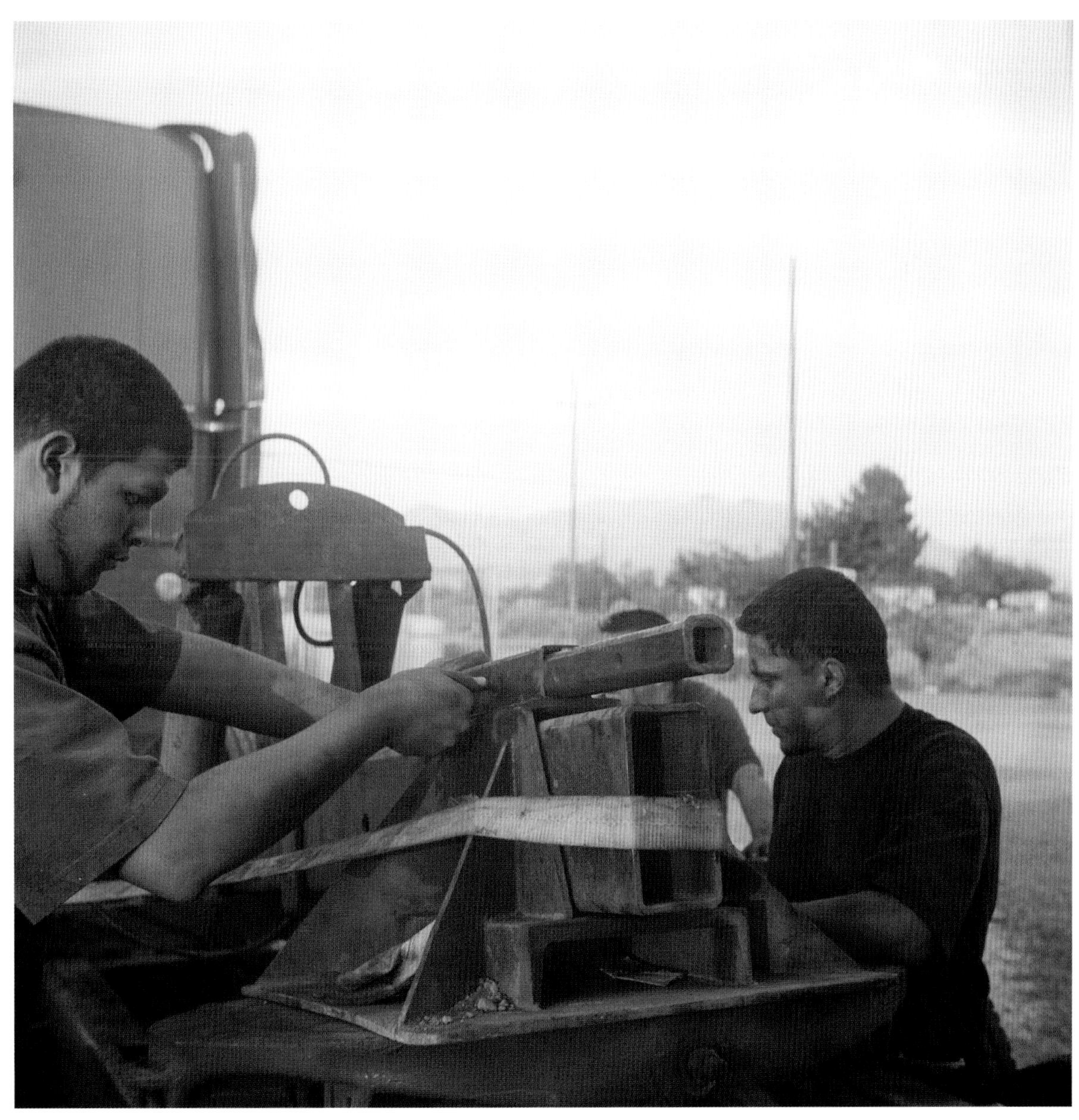

Port truckers preparing to tow a broken rig; I-15, Exit 12

Hog Express

Cows are shipped in the fall, spring and winter. Pigs are shipped in the summer, when the livestock season slows. The cows load easier. There are 206 pigs on this truck. They run through alleys into the two-level steel trailer equipped with air holes and sprinkler systems to keep them cool. "They'll get there by midnight, be slaughtered by 6 a.m.," says Adrian, 28, their driver. It's 95 degrees and the pigs are screeching; their smell carries over the baking asphalt. They fight to get closer to the walls of the trailer to put their snouts through the holes. Their skin is bright red like it has been rubbed with sand paper. Adrian picked them up in Alberta, Canada. He's watered them three times this trip, probably once more before he drops them in Modesto, California.

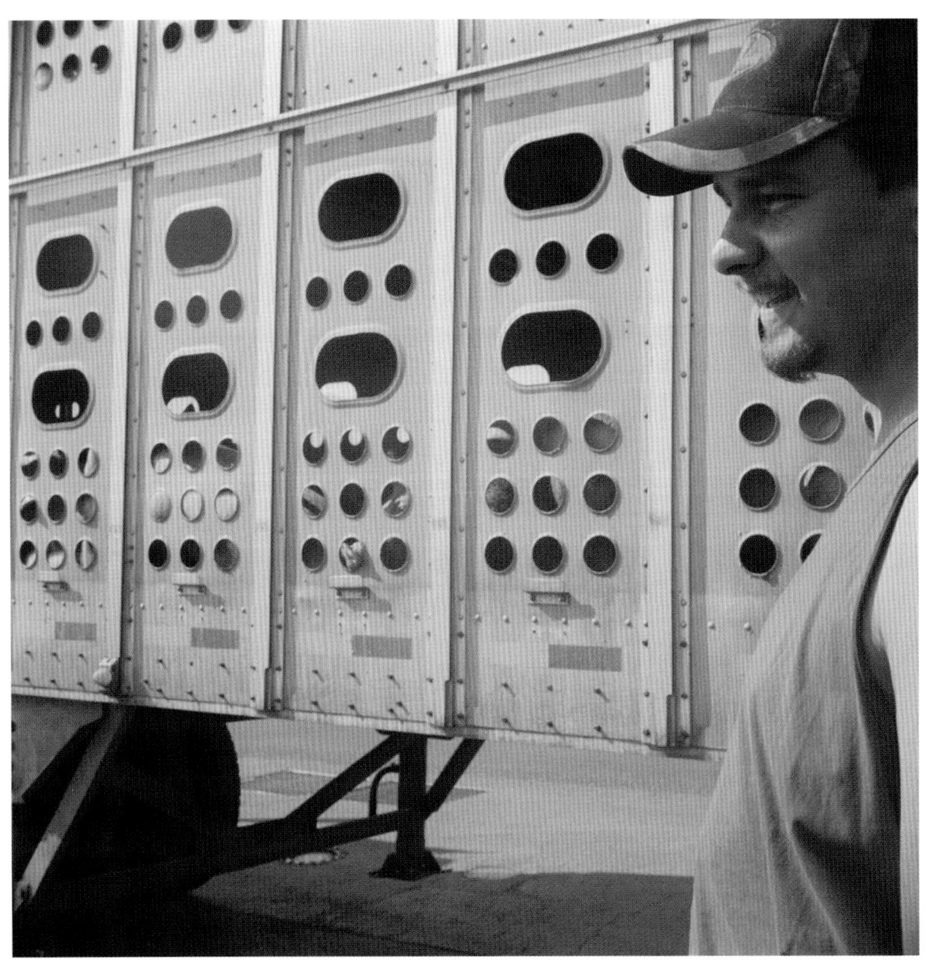

Adrian and hogs; I-80, Exit 176

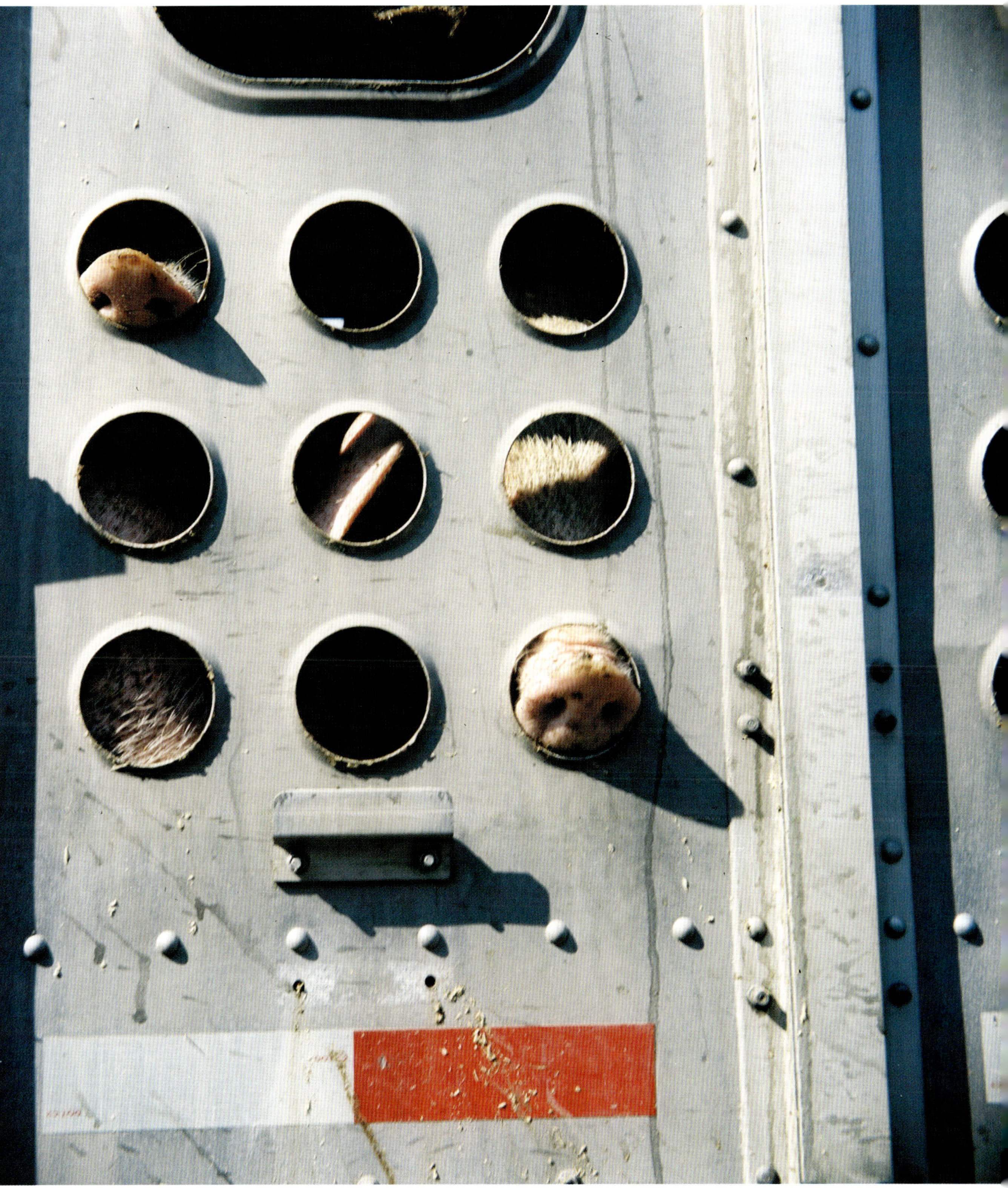

Water Warehouse

Jaime Albert is leaning on the counter. The girl behind the glass hands over the paperwork and the seal for his load of bottled water. "Do you know how long I've been here? A long time. I've been here since two. The reason I've waited all that time and I don't get paid for it is my dispatcher made the appointment with these ladies at seven in the evening." When he arrived, it was full of trucks and the warehouse was scheduled back-to-back with pick-ups. "It's like this every day." His trailer is at Gate 2. Men on forklifts quickly stack 20 pallets inside. "The axles cannot be over 34,000 pounds or you get tagged and you get fined. It could be in the thousands. The guys doing the loading, they know that in the front of my trailer, it's one single pallet and then maybe three or four rows of doubles and then another single pallet on the back." When they finish, he puts the serial coded plastic seal on the latch. An unbroken seal proves that nothing has been added to or taken off the trailer. When a load is considered valuable, the seal is steel-pin coded with a serial number.

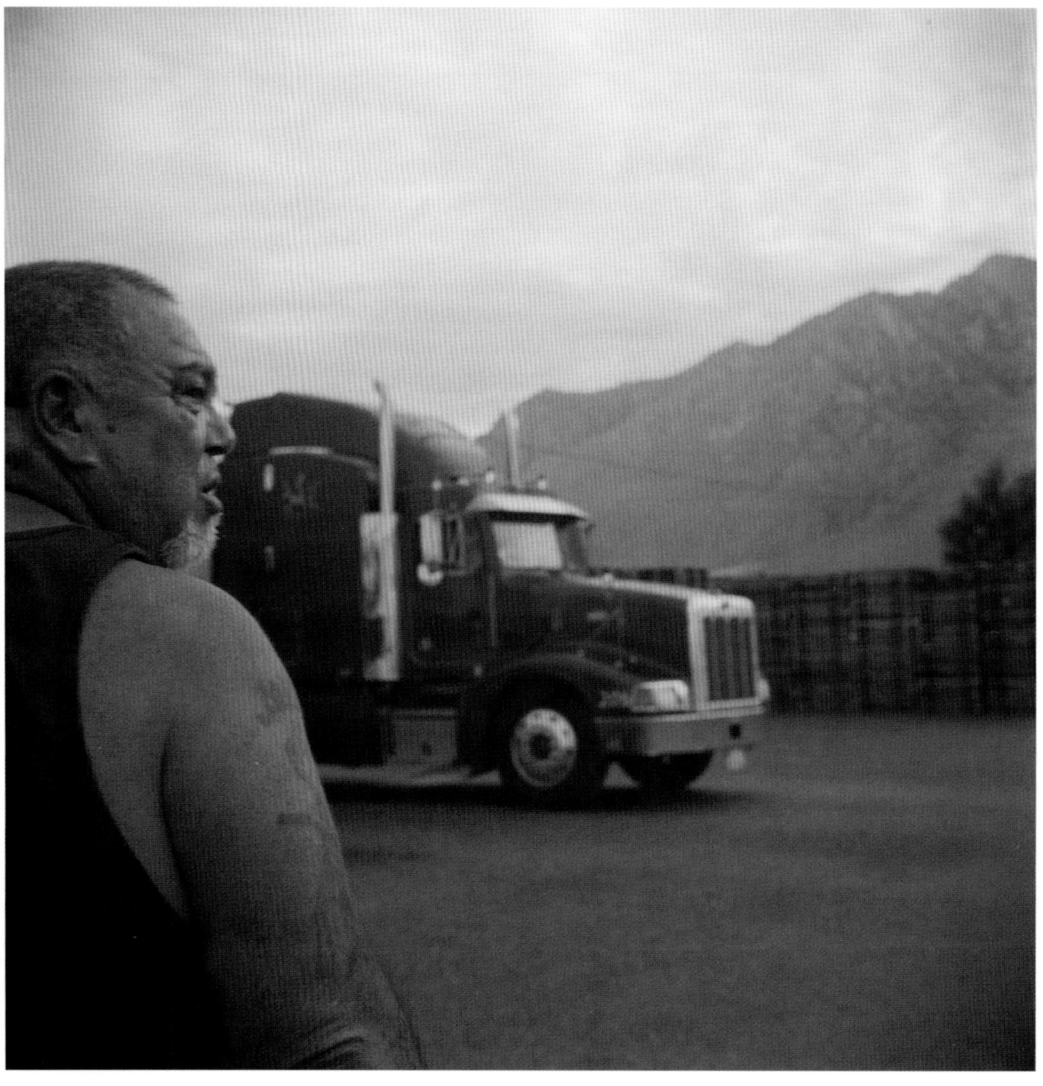

Jaime Albert, picking up water in Calistoga, California

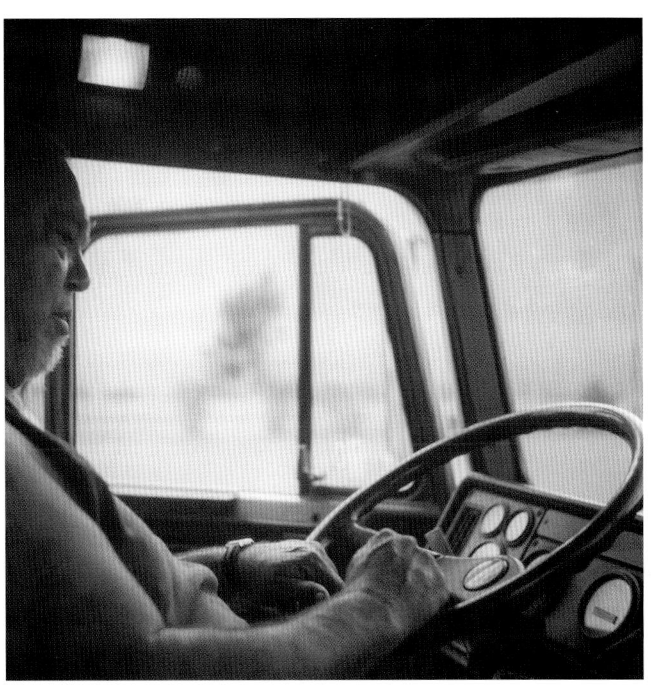

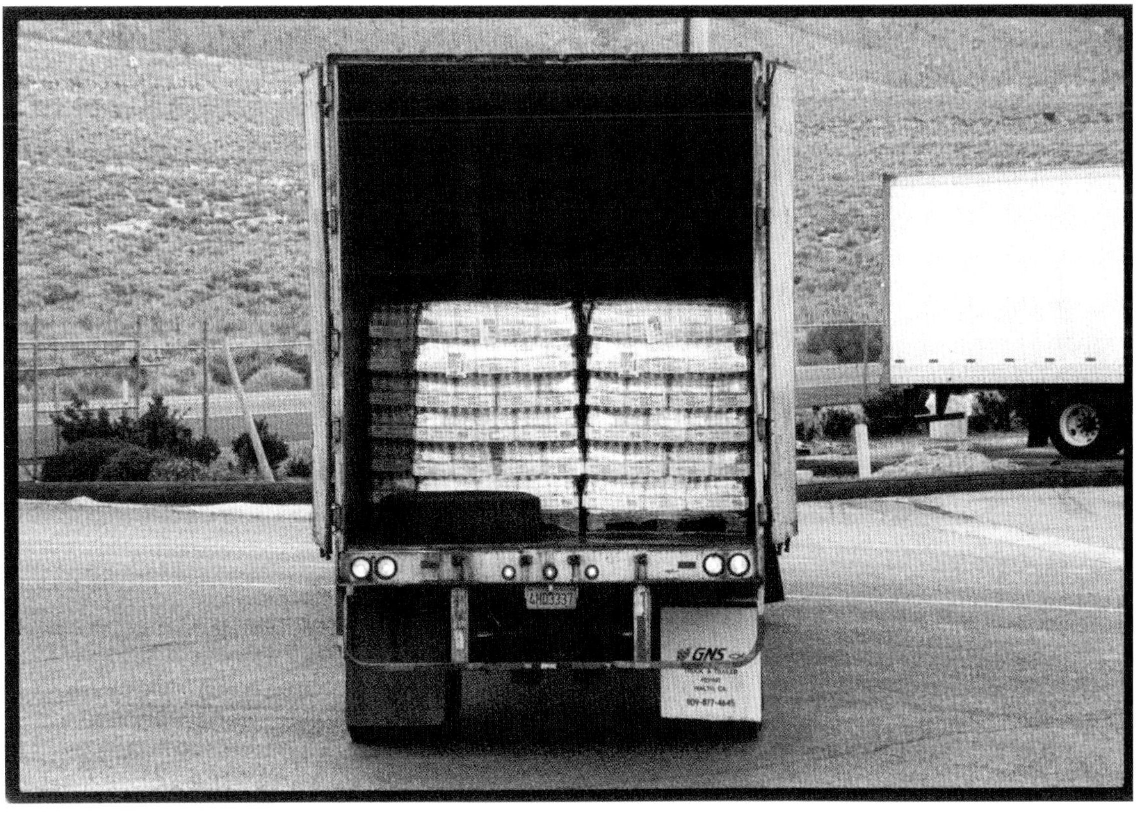

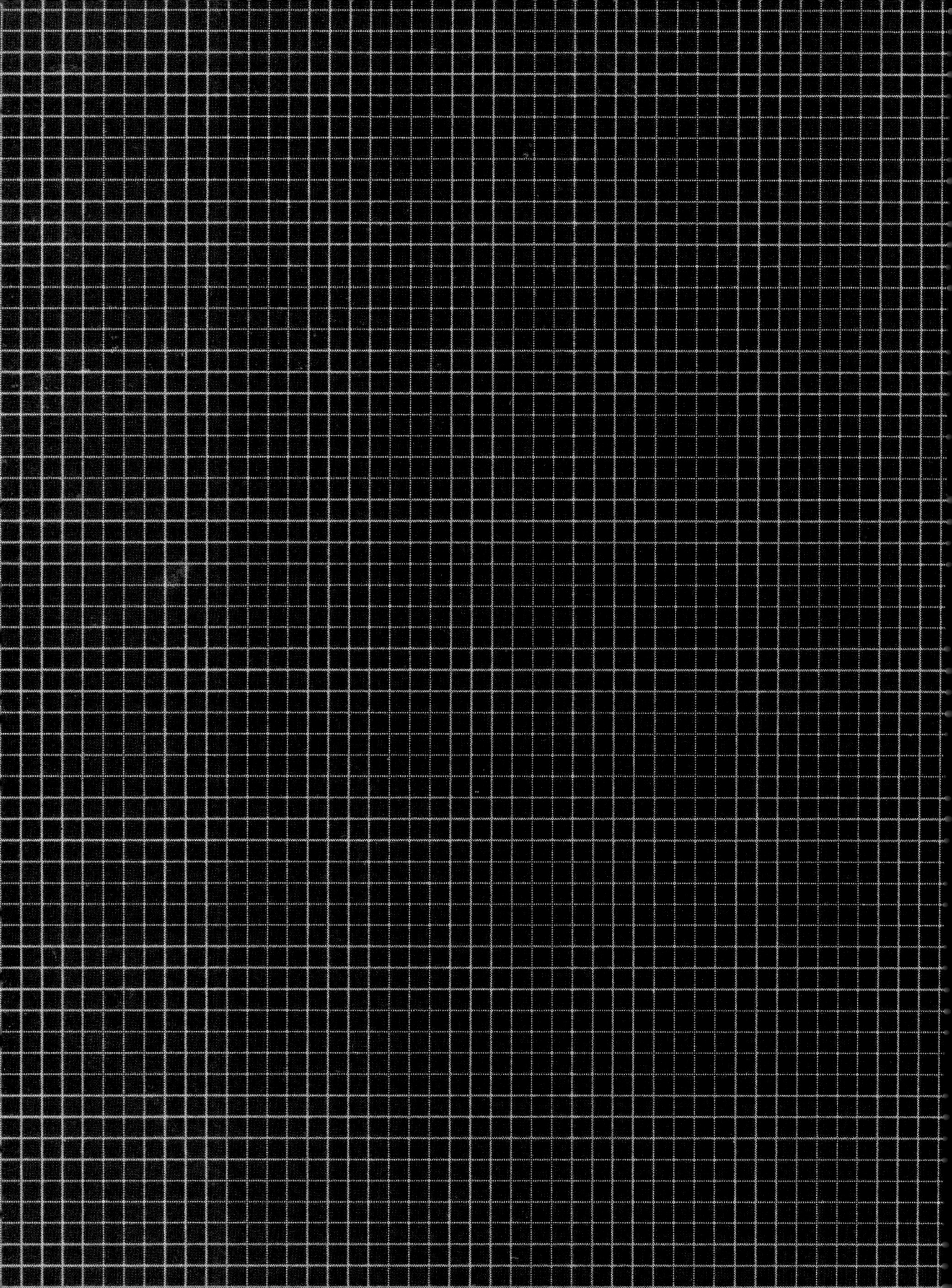

PART FOUR
Health, mental and physical

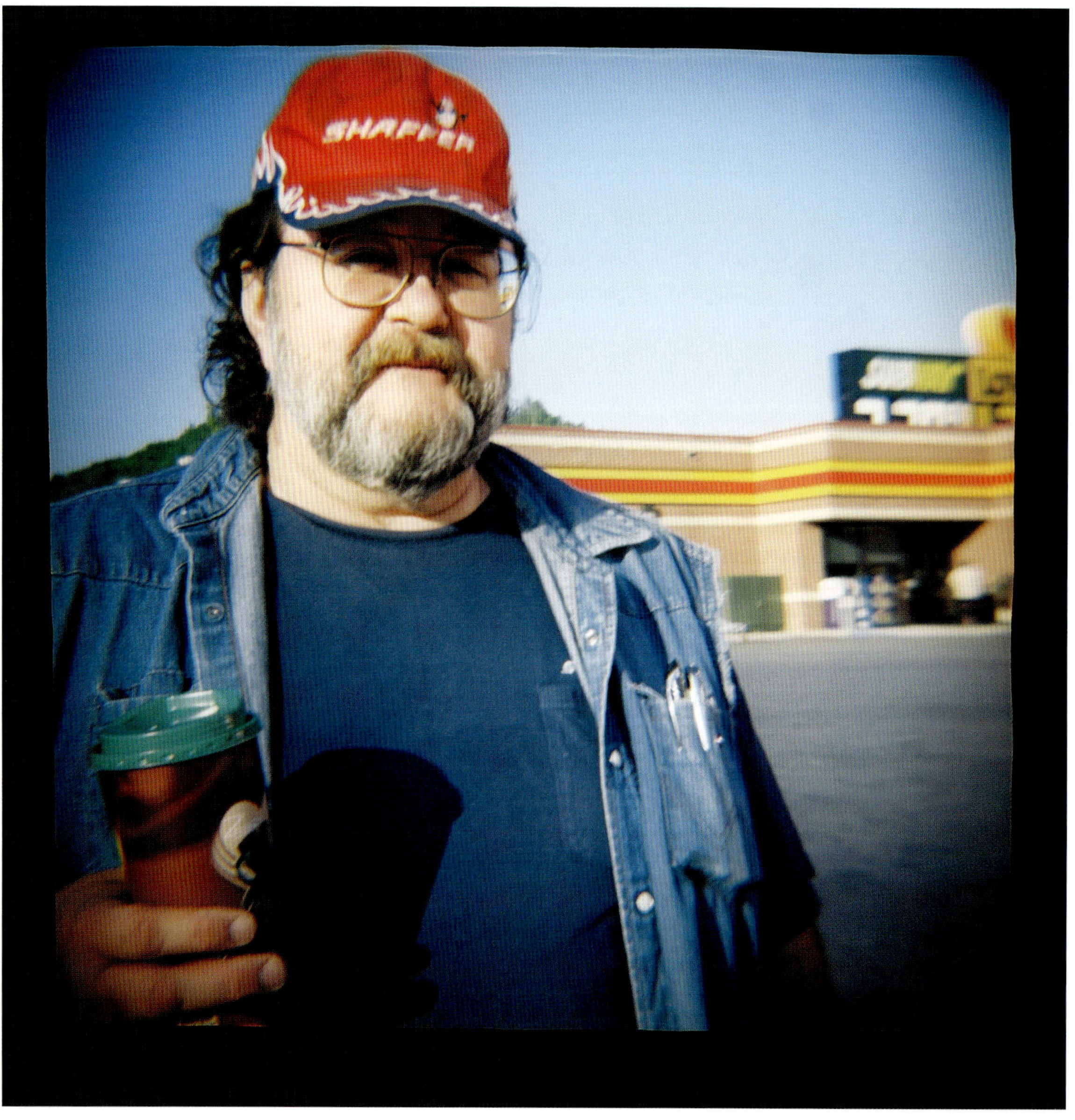

Ron Edwards, having coffee before his overnight driving shift

"I wanted to be a scientist, but my grades were not good enough. I wasn't in good enough shape to go into the military."

Fatigue (def.)

Diminished alertness, lack of energy, drowsiness; caused by inadequate sleep, interrupted circadian rhythm.

Common solutions: caffeine, sugar, nicotine, Monster energy drink, All Day Energy liquid shot, Black Ice, Stacker 3, Ephrine, Ginseng Blast, Yohimbe Rush, Manis, methamphetamine.

Common side effects: insomnia, high blood pressure, hypertension, heart attack, stroke.

Side effects (of side effects): fatigue, highway crash, death, seizure, explosion.

A Nine-Foot Drop From The Cab

Common occupational hazards: chemical leaks, lung cancer, depression, loneliness, seizures, explosions, back injury, exposure to diesel fumes, broken families, highway crashes, heart attack, obesity, black-outs, sleep apnea, high cholesterol, hearing loss.

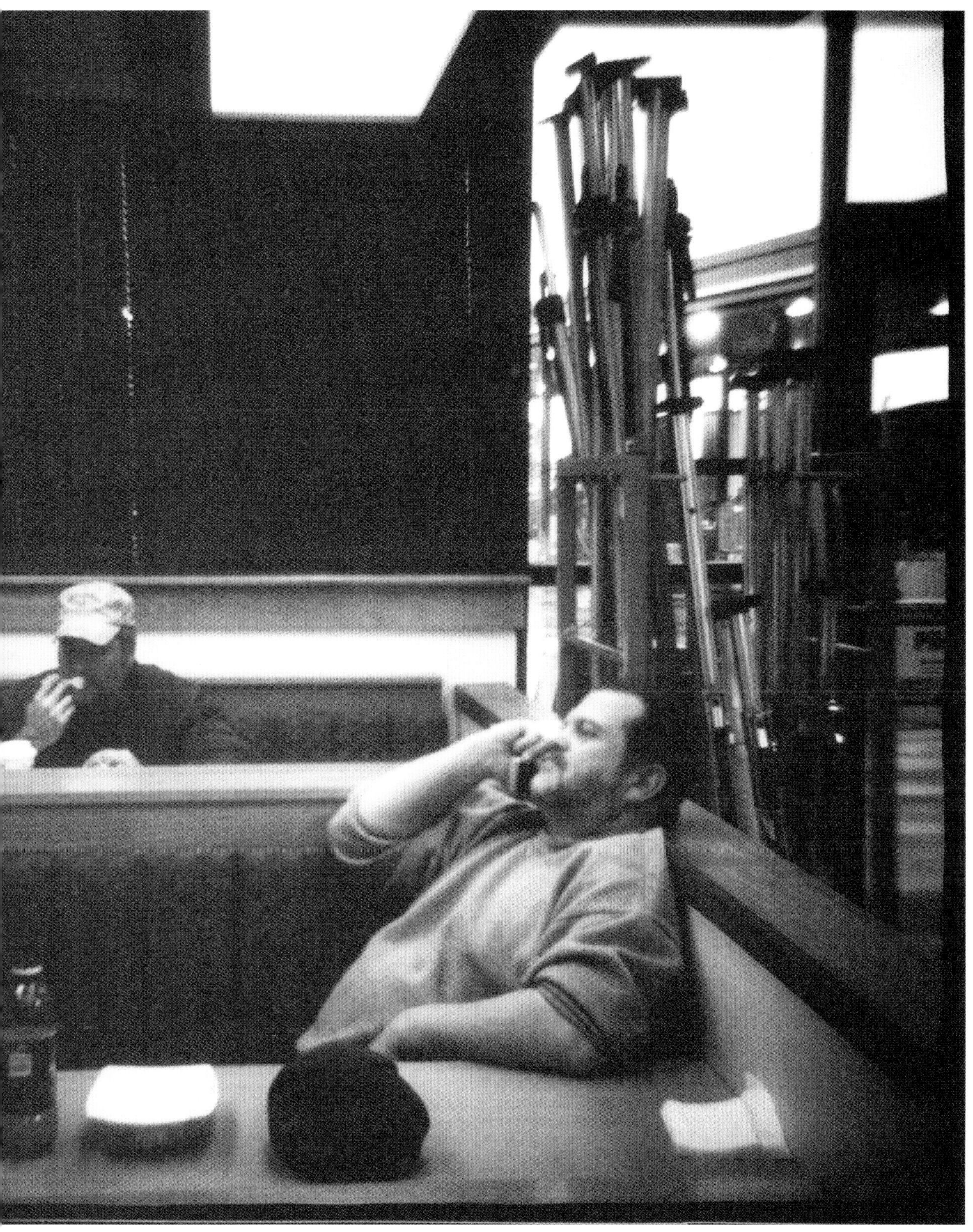

Stan Ruth, making a phone call inside of a truck stop

Family: married, 2 girls, 2 step-sons
Previous job: supervisor in extrusion plant, 20 years

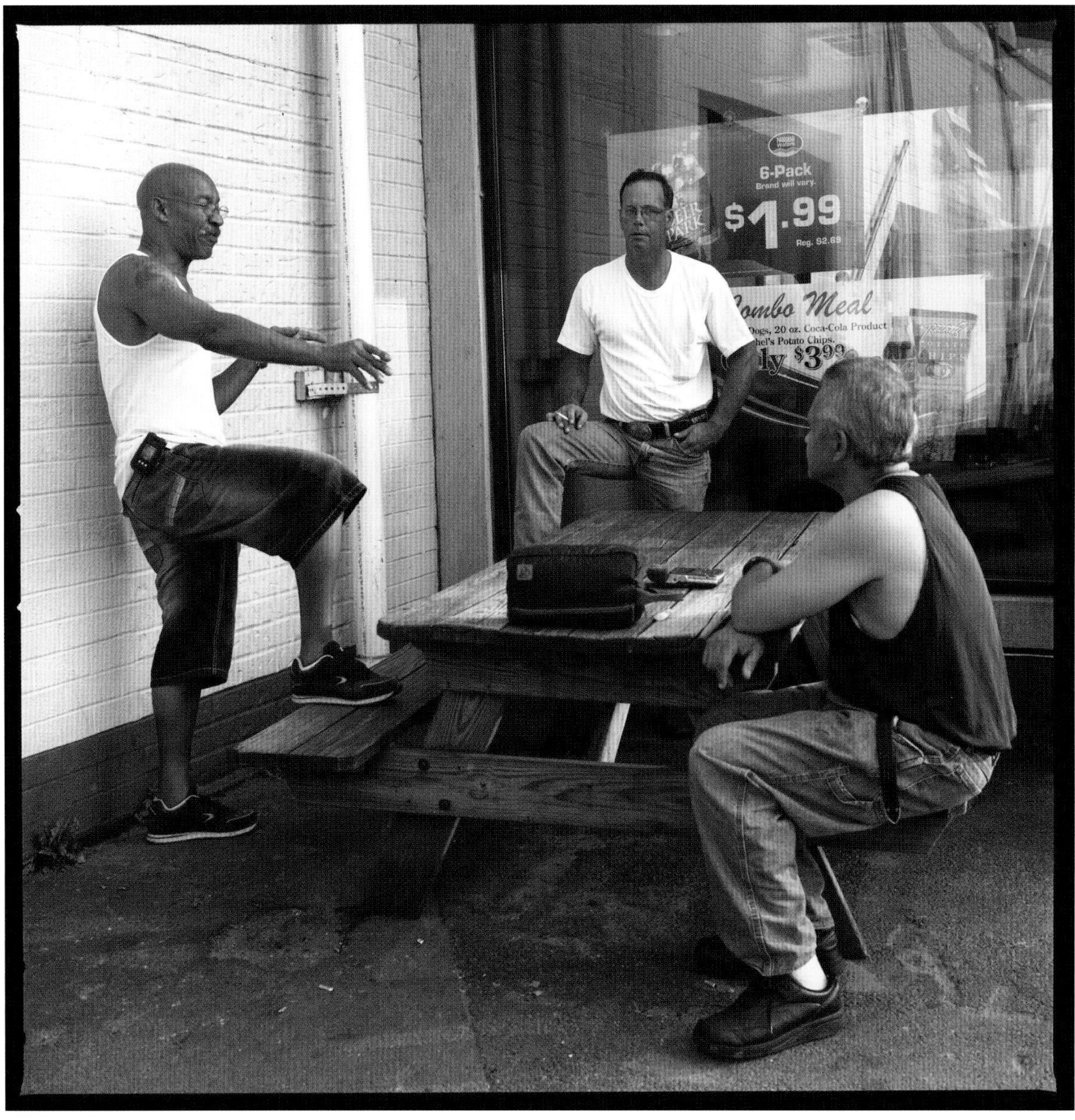

Otis Dixon, talking with Mike Nieves
and Floyd Mallory; I-80, Exit 4

Sleeping and Eating

For sleeping, truckers usually park in one of three places: truck stops, rest areas, Interstate ramps. A ticket for sleeping on a ramp costs between $100 and $180. Restful sleep in a truck is an acquired skill. Otis Dixon explains, "I sleep better in my truck than in my bed at home. To hear the noise of the trucks around me is like a comfort. I go home and I toss and turn for the first day or two. If I don't hear no noise or nothing, I wake up and try to figure out where I'm at." He says that eating on the road is also an acquired skill. "You see a restaurant you might like to go eat at, but you can't park, so you can't go eat at that restaurant, and you're stuck with truck stop slop. You get tired of eating Subway, so then you go into a McDonald's kick. You get tired of eating McDonald's and you get into a Wendy's kick. Then when you get tired of all that, you go back into the truck stop."

Shower No. 3

Before I did this, I was a guard at a maximum security prison for 12 years. There was at least human contact. Now look at who is shelled inside a box. Truckers sit for twice as long as office workers, ears ringing, hearing shot from the sound of the engine running and the trailer rattling. Never a kindness, soft skin. I cannot hear a whisper if there were one. Wispy dry hair and the sun is going down. I watch from the window inside, waiting for a shower to open up. Coffee, wash off the coffee. It is dirty in my head. They want me to pick up another load 500 miles away before I head home. Eyes blurry, squinting all the time, the pee bottle spilled on the floor this morning. Can't shit right anymore after holding it in so many times.

[An automated voice] *Shower number three is now available. Shower number three is now available.*

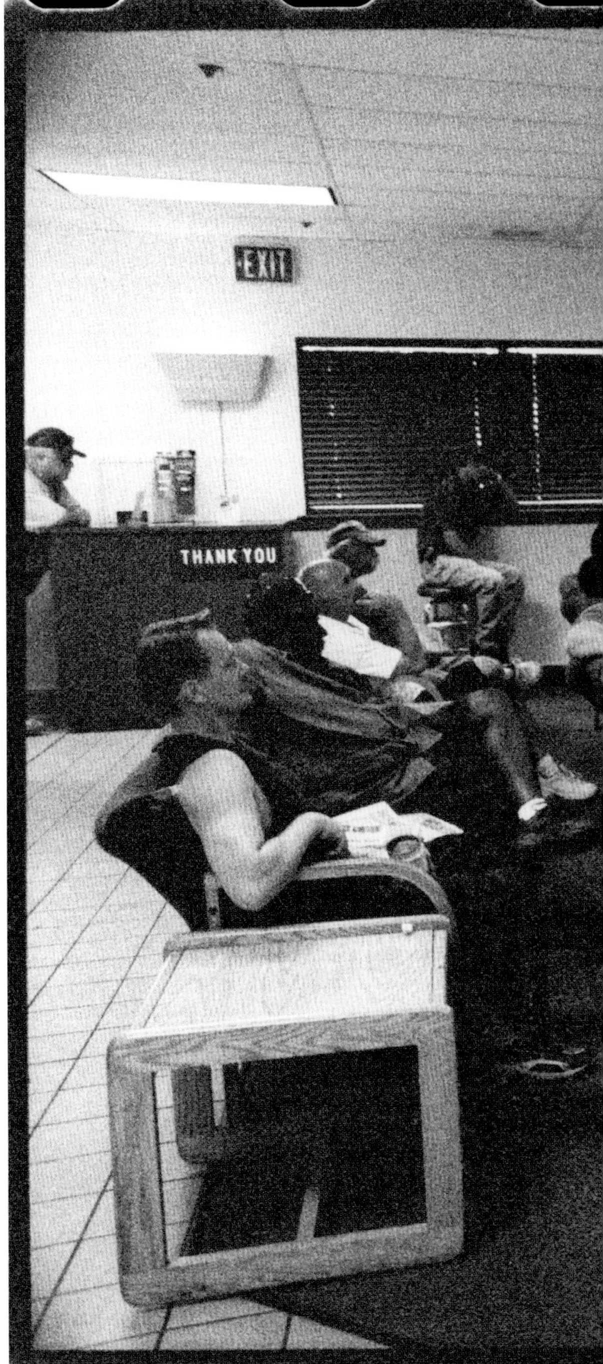

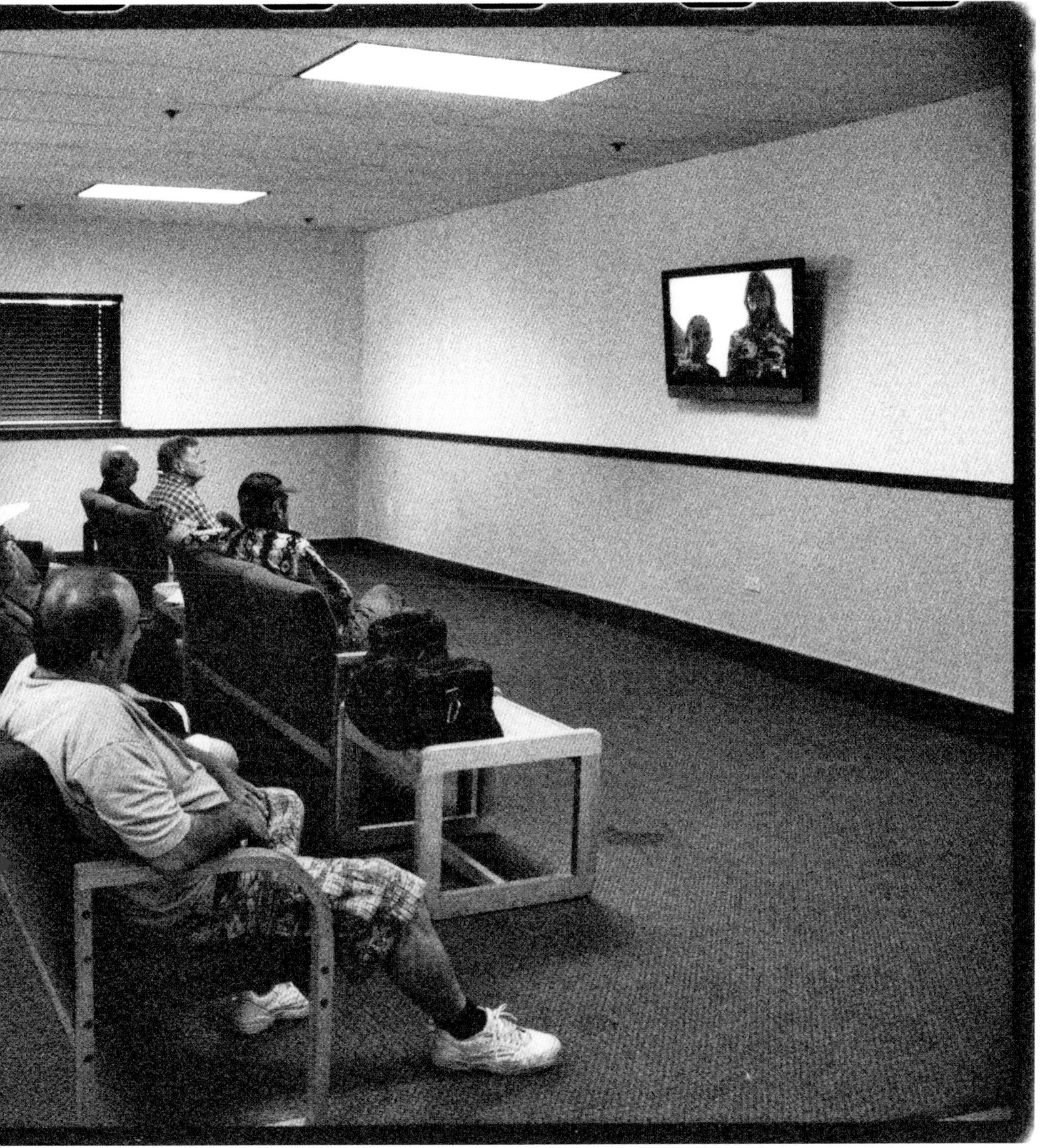

Truck stop lounge

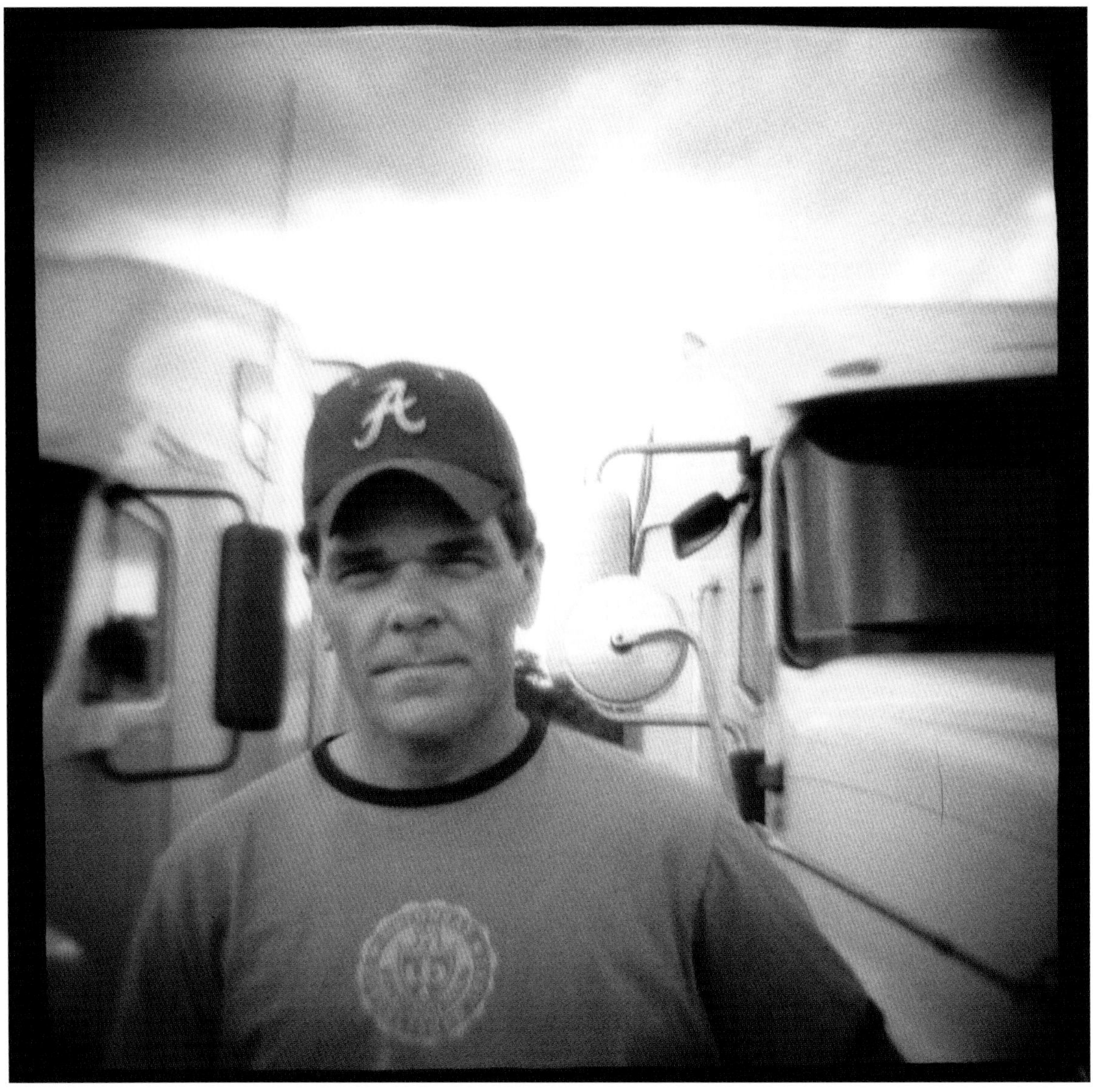

Waking from a nap

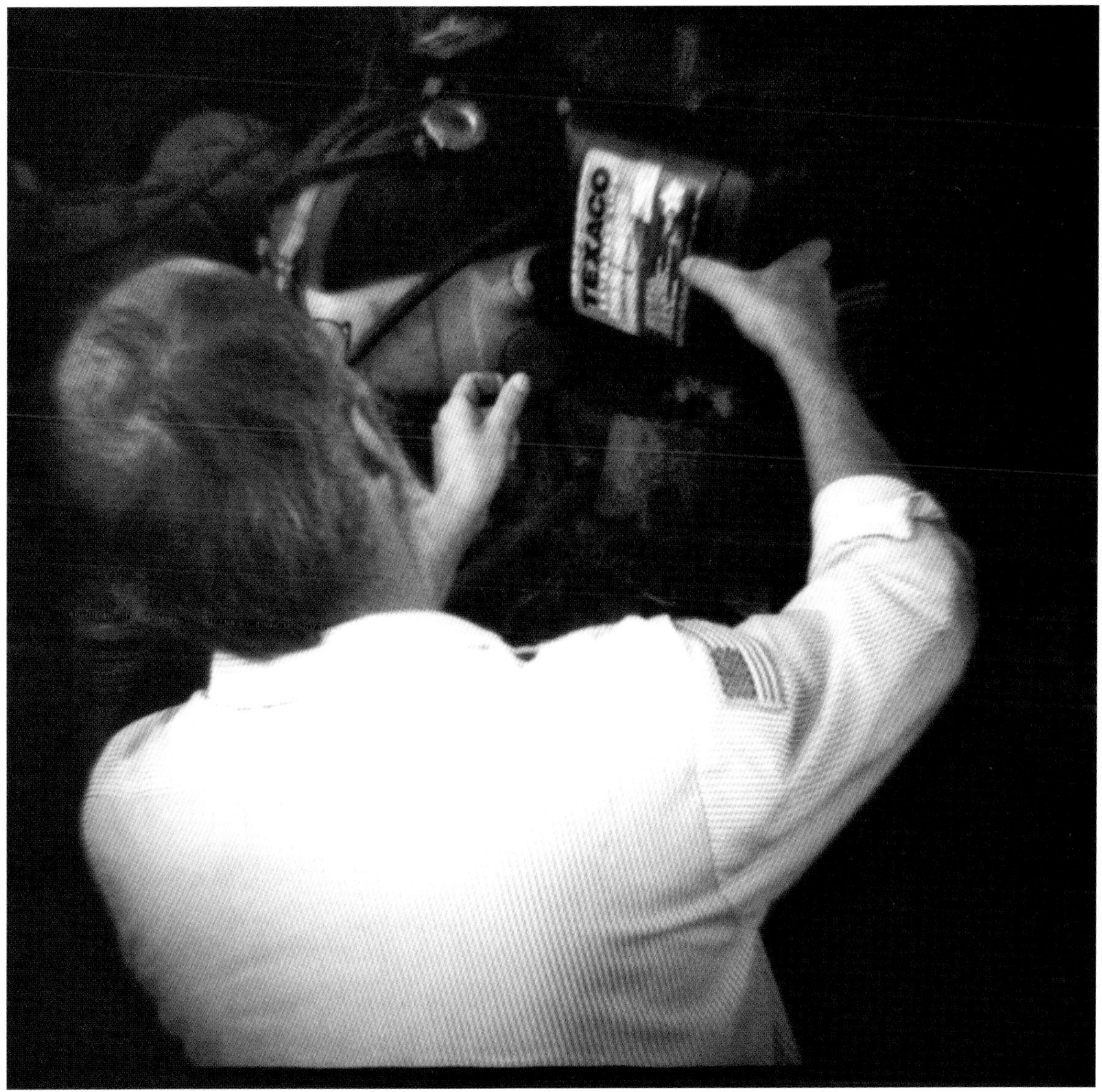

Oiling the engine

Runaway Truck Ramp (def.)

A long lane filled with sand or gravel usually located on steep grades, like mountainous areas, where truck brakes can overheat and fail; usually appear before a change in road curvature or before an intersection in a populated area.

1. Arrester bed: a gravel-filled ramp that uses rolling resistance to stop the vehicle.
2. Gravity escape ramp: a long upwardly-inclined path.
3. Sand pile escape ramp: a short length of loosely piled sand.

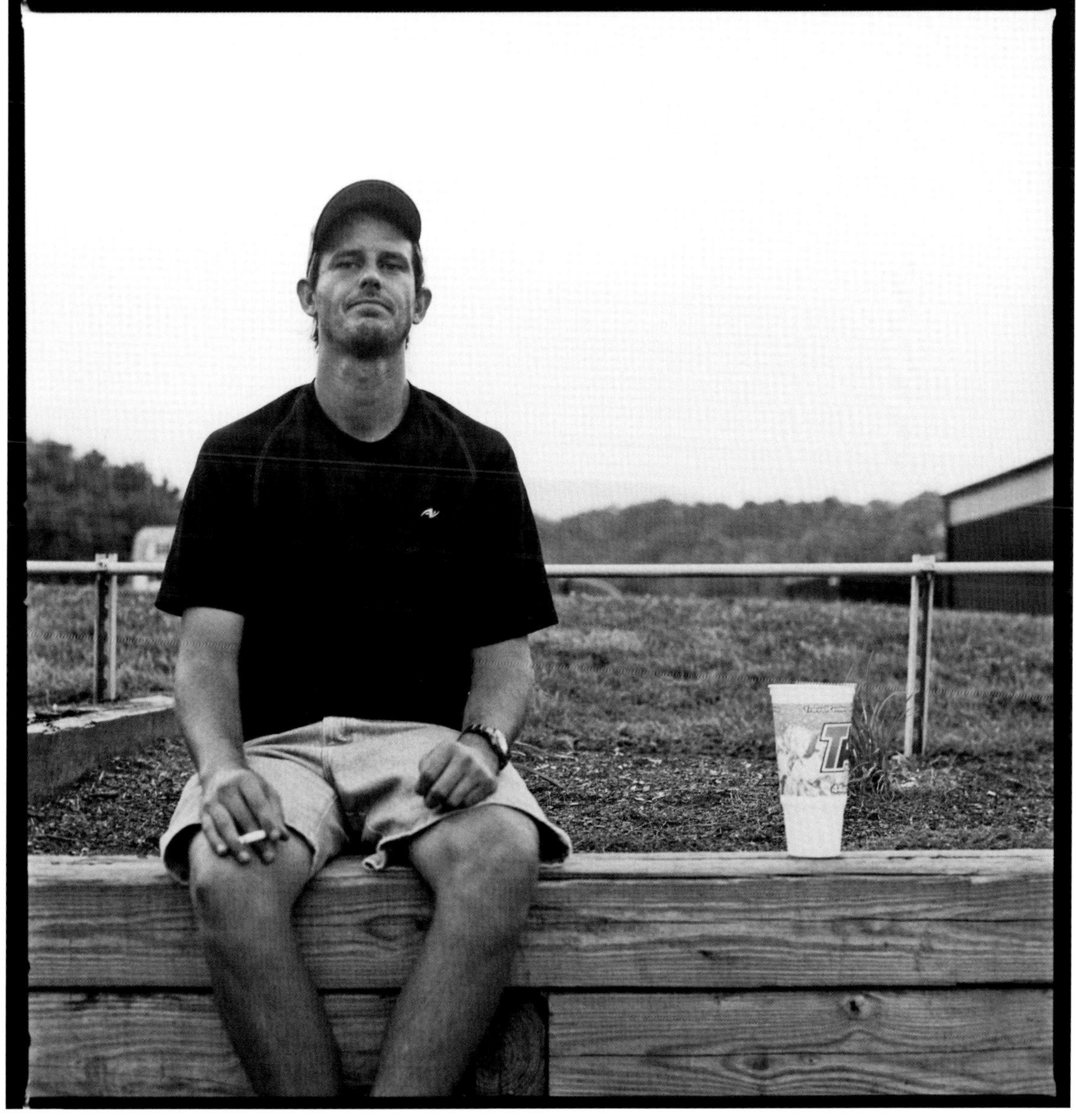

Brett Doyscher, a trucker and German immigrant; I-80, Exit 4

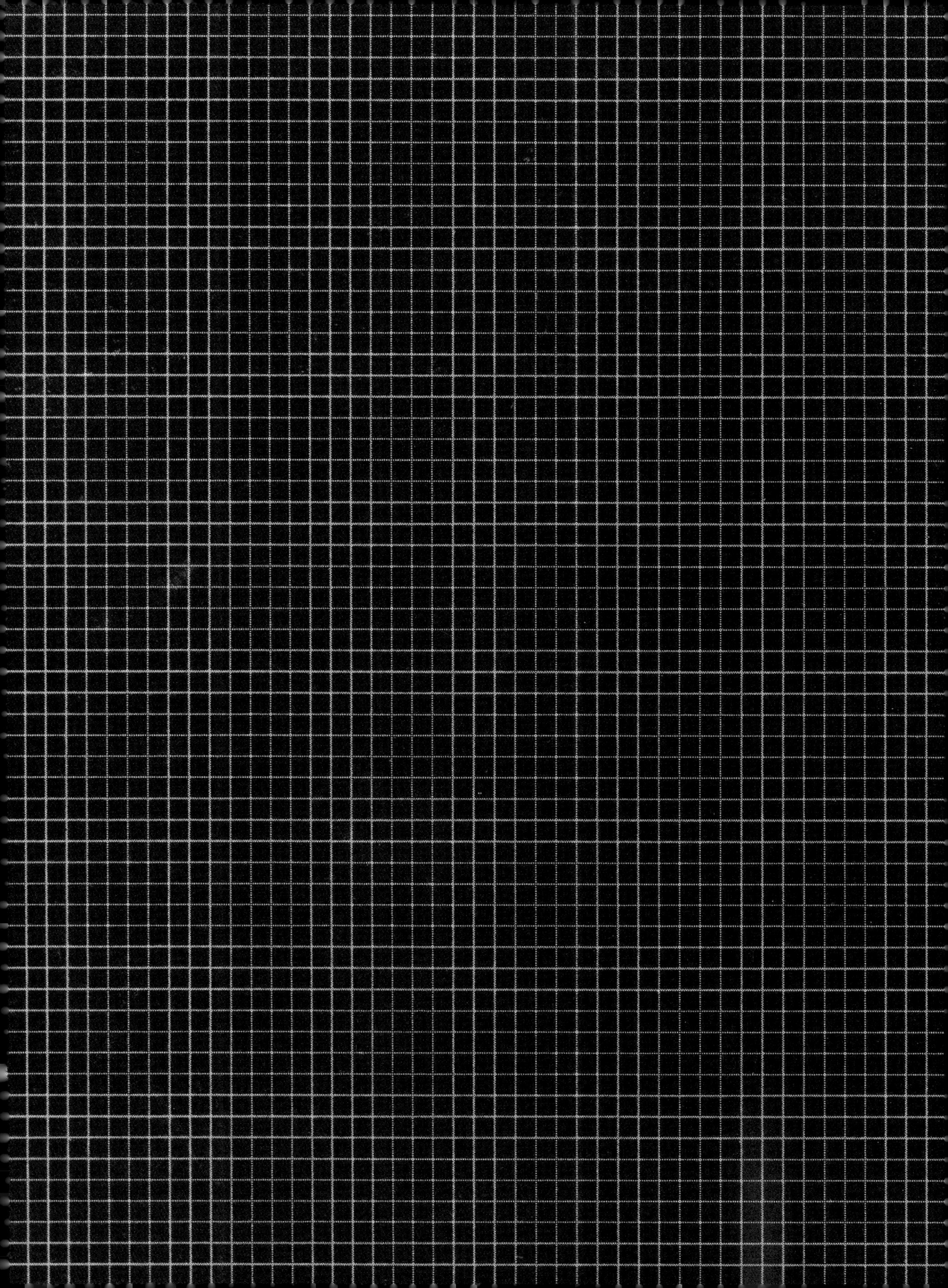

PART FIVE
Technology, freedom, family

For Recruiting, Press One Now.

"You call any recruiter for any of these companies and they're going to feed you a line of bullshit, see the country and see this, see that. The only thing you're going to see is a sign that says this is where such and such is. Because you're not going to get there with a big truck and you don't have time anyhow to worry about going. The time you got is to sleep." – Floyd Mallory, company driver

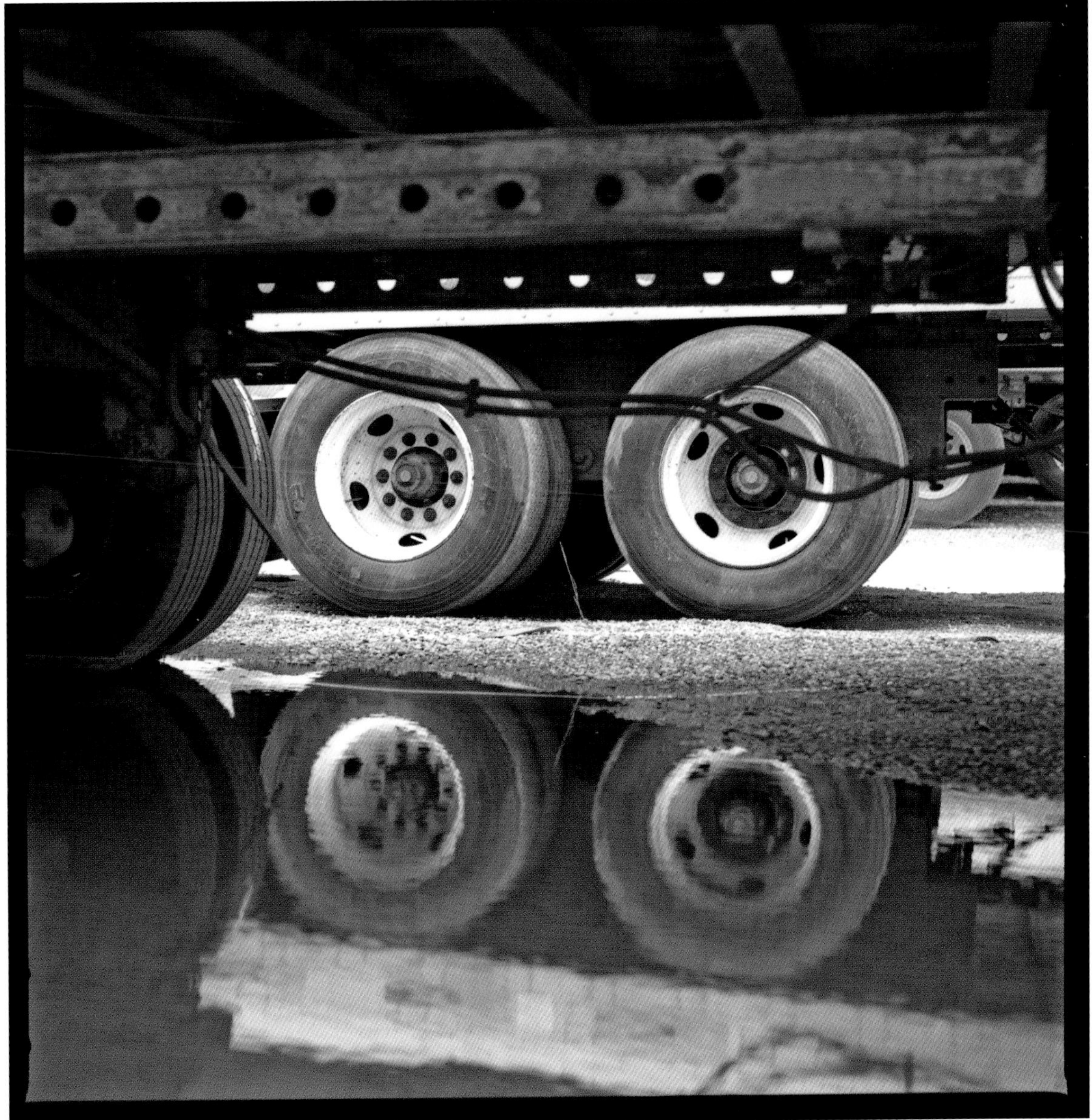

Tires reflecting in puddle

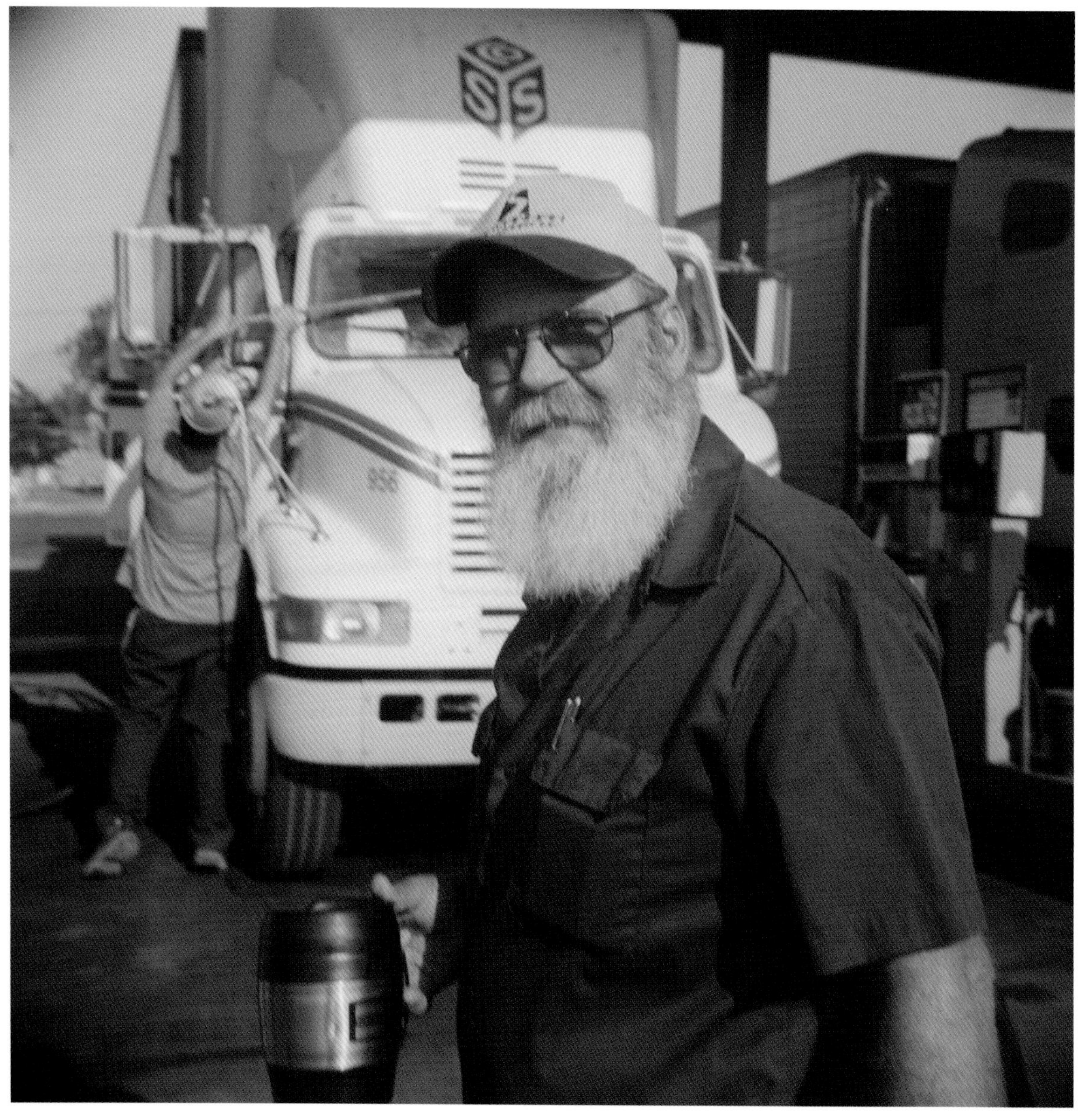

I-80, Exit 176

Shrinking Landscape

Jeans and cut off sleeves move steadily through fuel islands. A woman is leaning against the brick wall, watching them. Long grey wild curls frame her face, cover her shoulders and run down her back. "I'm a rider, not a driver," she says, looking over her coffee, eyeing potential drivers. She's been 50,000 miles in a month and a half. "My rule is I sleep in the front." A couple hundred yards beyond her gaze, quality time between a man and his grandson is pressurized—the strawberries are late, bound for Nebraska. Further on, an older couple and their two young grandsons enjoy a summer trucking vacation; they share buffet dinner and the same blue eyes from North Carolina. The wrinkled clerk's eyes dart back and forth at the revolving door, regretting no time for smoke or chat with the off-duty clerk across the counter.

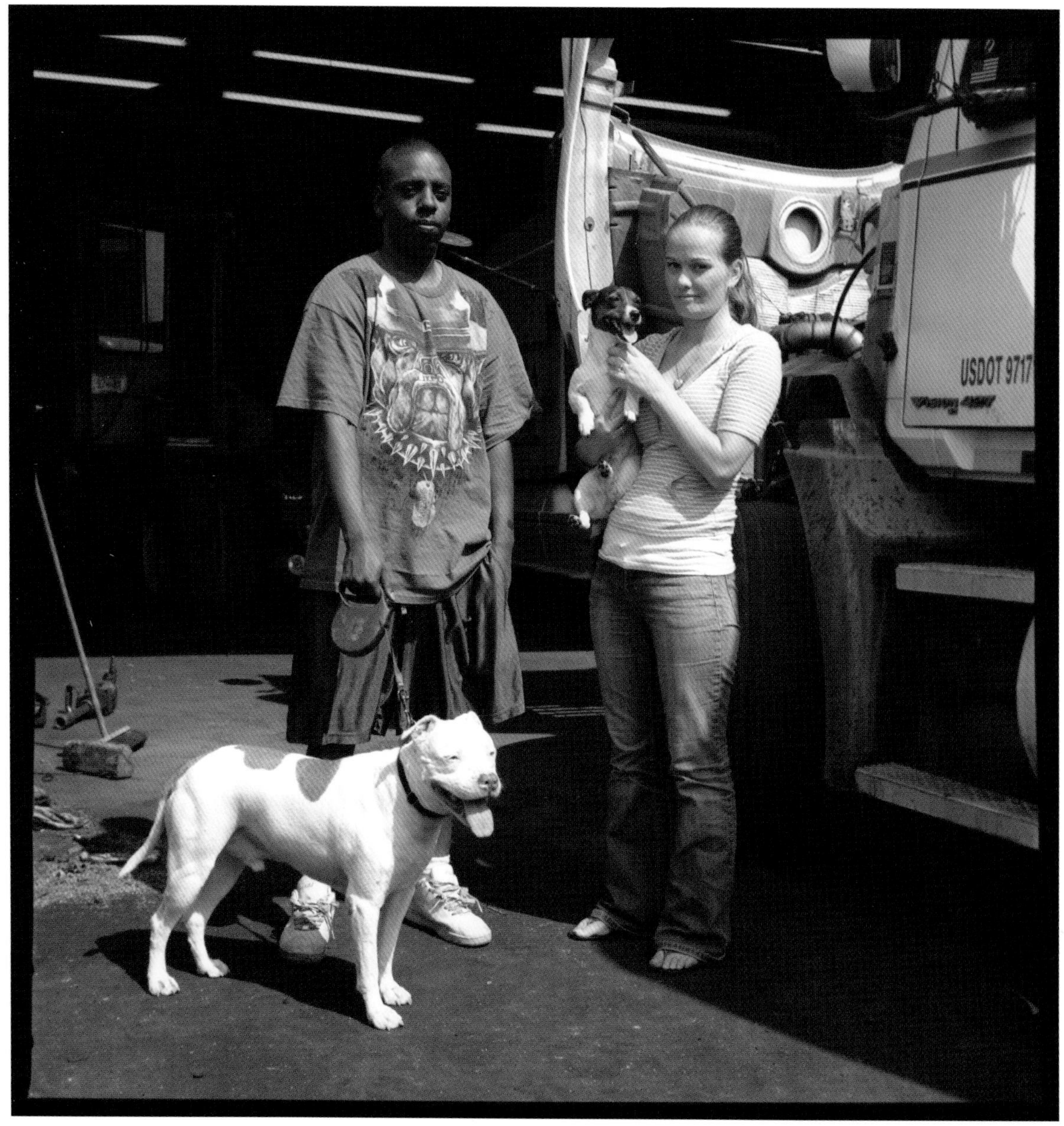

Arthur E. White, Jr, Marlena Lasure,
Hannibal and Sassy

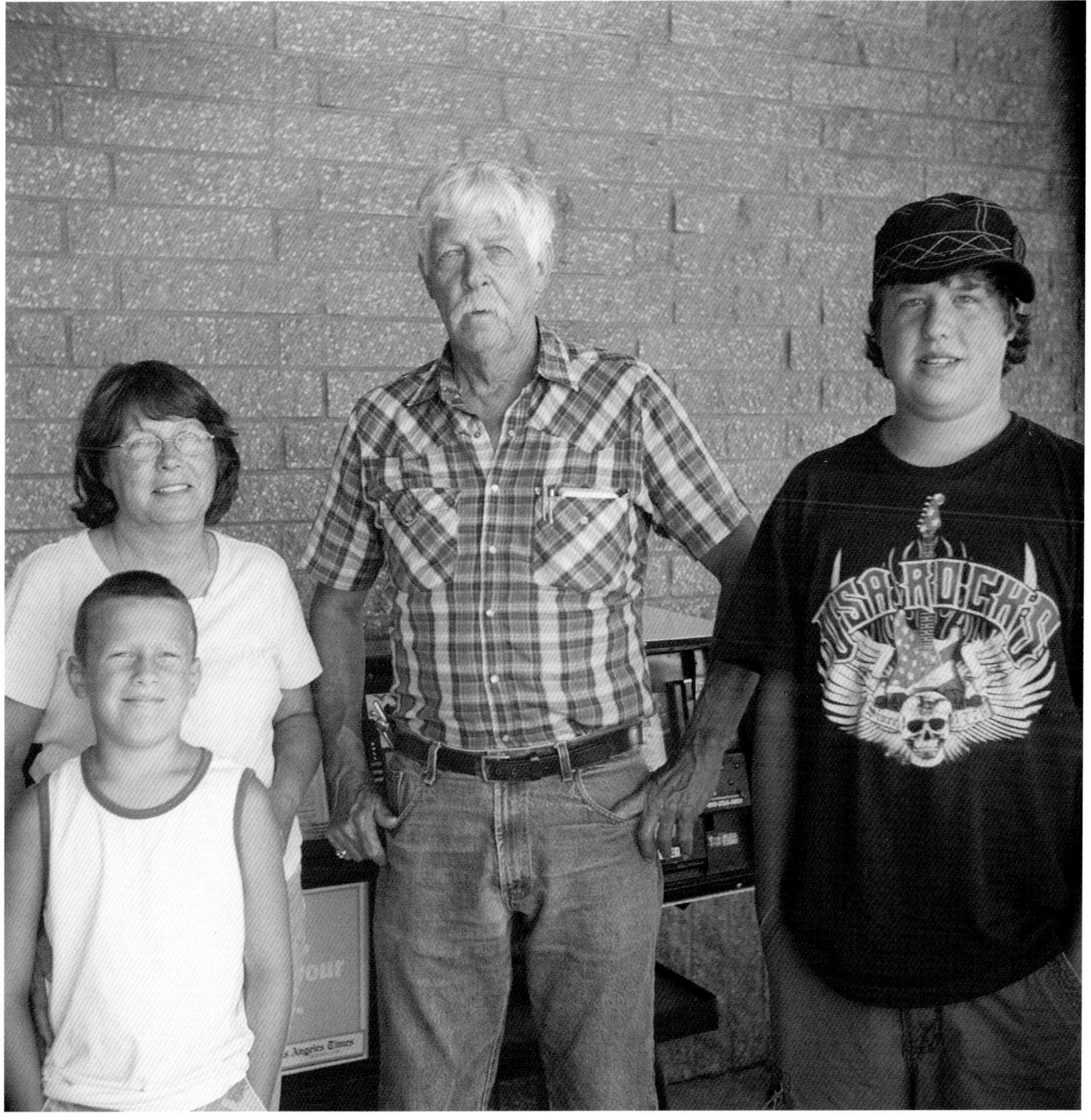

Summer ride along: husband and wife team with their two grandsons

80,000LBS OF FREEDOM OR CONVERSATION BETWEEN TWO COMPANY DRIVERS AND AN OWNER-OPERATOR.

OWNER-OPERATOR: I'm working for myself because I lease my truck.

DRIVER #1: I don't see much difference between a company driver and a lease driver.

OWNER-OPERATOR: Depending on the company, it's different.

DRIVER #1: The lease driver is more handcuffed. He has to pay that lease. He's got to work more.

OWNER-OPERATOR: I'm not more handcuffed. Are you forced dispatch, or can you tell them, "No I'm not going here, I'm not going there"?

DRIVER #1: Basically, if you don't go there, you might not even get dispatched to certain areas. In a way, they force you.

OWNER-OPERATOR: I get load offers sent to me. I take and choose what load I want. If I don't want any of them, I don't have to take them. If I want to go home, I can go home. And I don't have anybody telling me that I can't. As long as I've got money to put fuel in that tank, I can go wherever I damn well please.

DRIVER #2: I had that freedom. When it comes time to pay them bills, that freedom becomes confining because you've got to get out there to pay for that truck. Now, after you pay for the truck, you've got a little more freedom, but now you've got a choice of how much you want to take home. I had it figured out that it took me 3,400 miles to pay my bills. Anything after 3,400, that's me. So I'd run harder after 3,400 because now we talking about what I'm taking to the house.

DRIVER #1: And you hoping you got a good broker and a good company to support you.

OWNER-OPERATOR: Yeah, it all comes down to that.

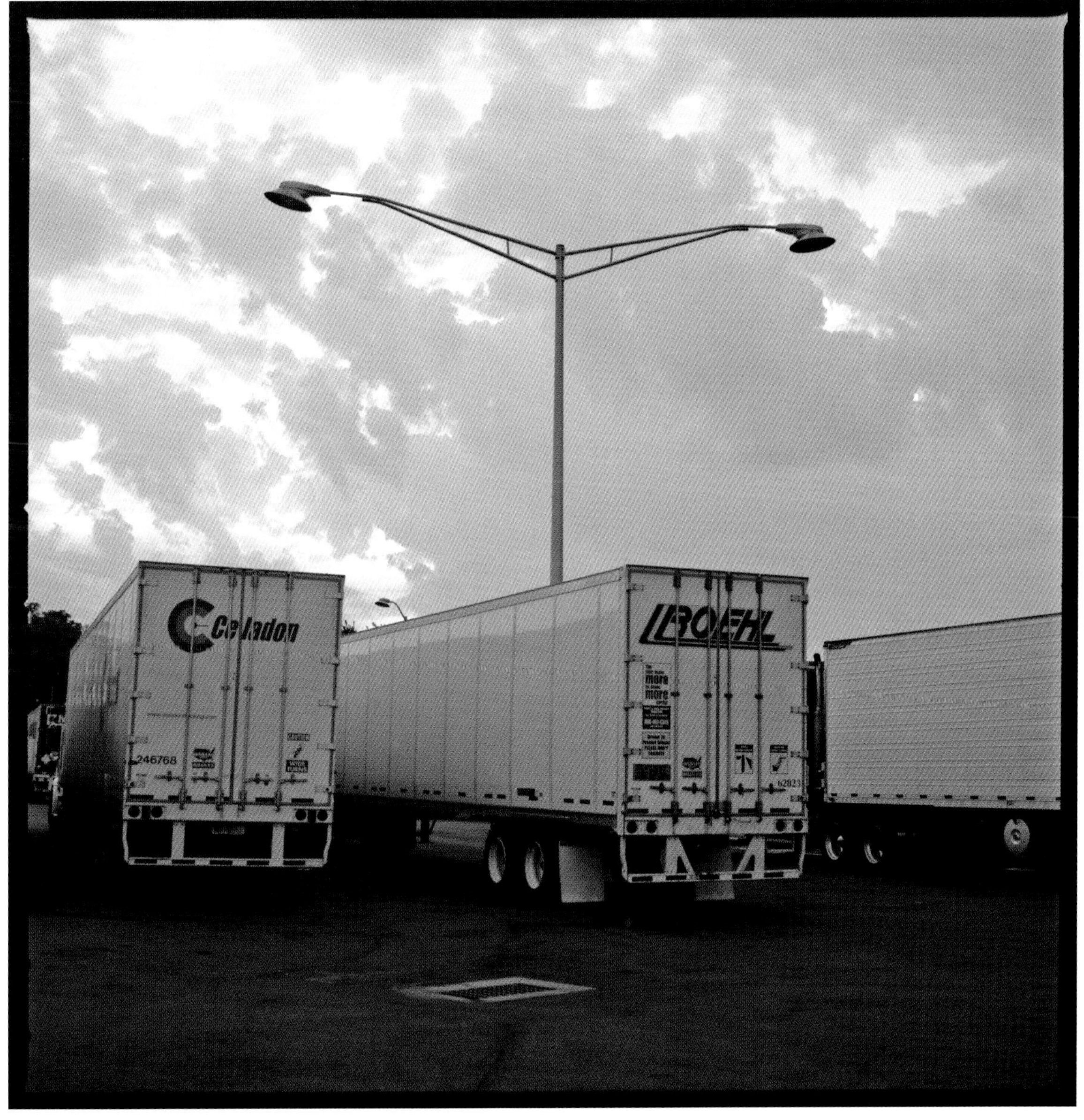

Three trucks, Modena Service Area;
I-87, mile marker 65

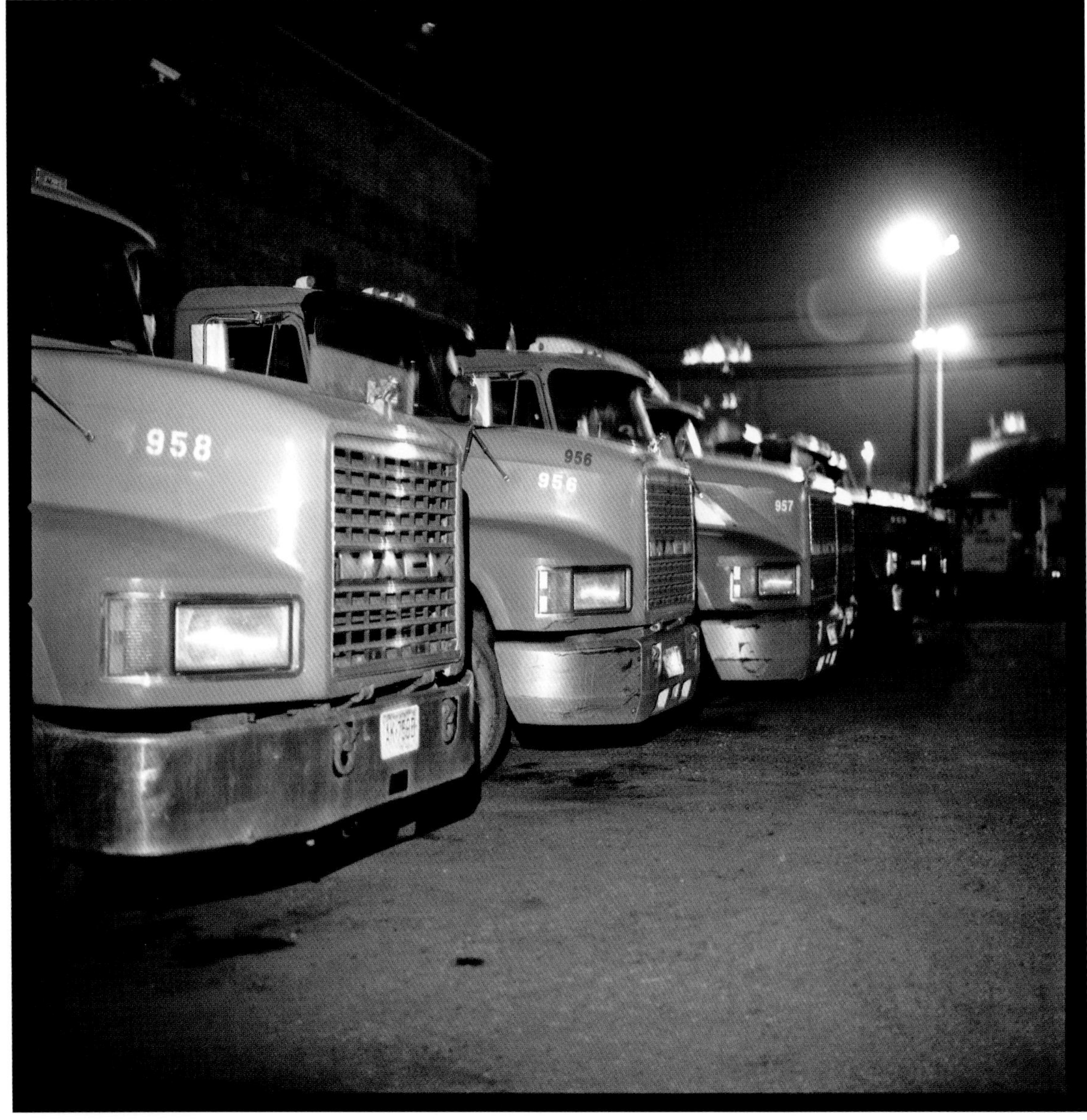

958, 956 and 957, Mecca & Sons;
Jersey City, New Jersey

Holdover

Qualcomm (DEF.) A device that let's them know where you are through an onboard computer.

His name is The Snowman. His previous job was Vietnam and cocaine. "I don't do drugs. Thank God I got off that crap in '75. That's why I got the handle Snowman." He's wearing a padded plaid jacket with a quilted nylon lining, impressing upon an aloof and fume-soaked winter the sentimentalities of trucking in it's golden age. Life was in the palm of his hands, tearing through a young, unclogged Interstate, unattached and flying. His real name is Ray McMillan. Behind him is an ad for Bluetooth CB radios. Payphones are a rarity and The Snowman is now tracked by satellite. His route and speed are provided to his company by Qualcomm. An automated voice from his GPS device gives him driving directions. He communicates with his dispatcher through an onboard computer on where to pick up, drop off and fuel up.

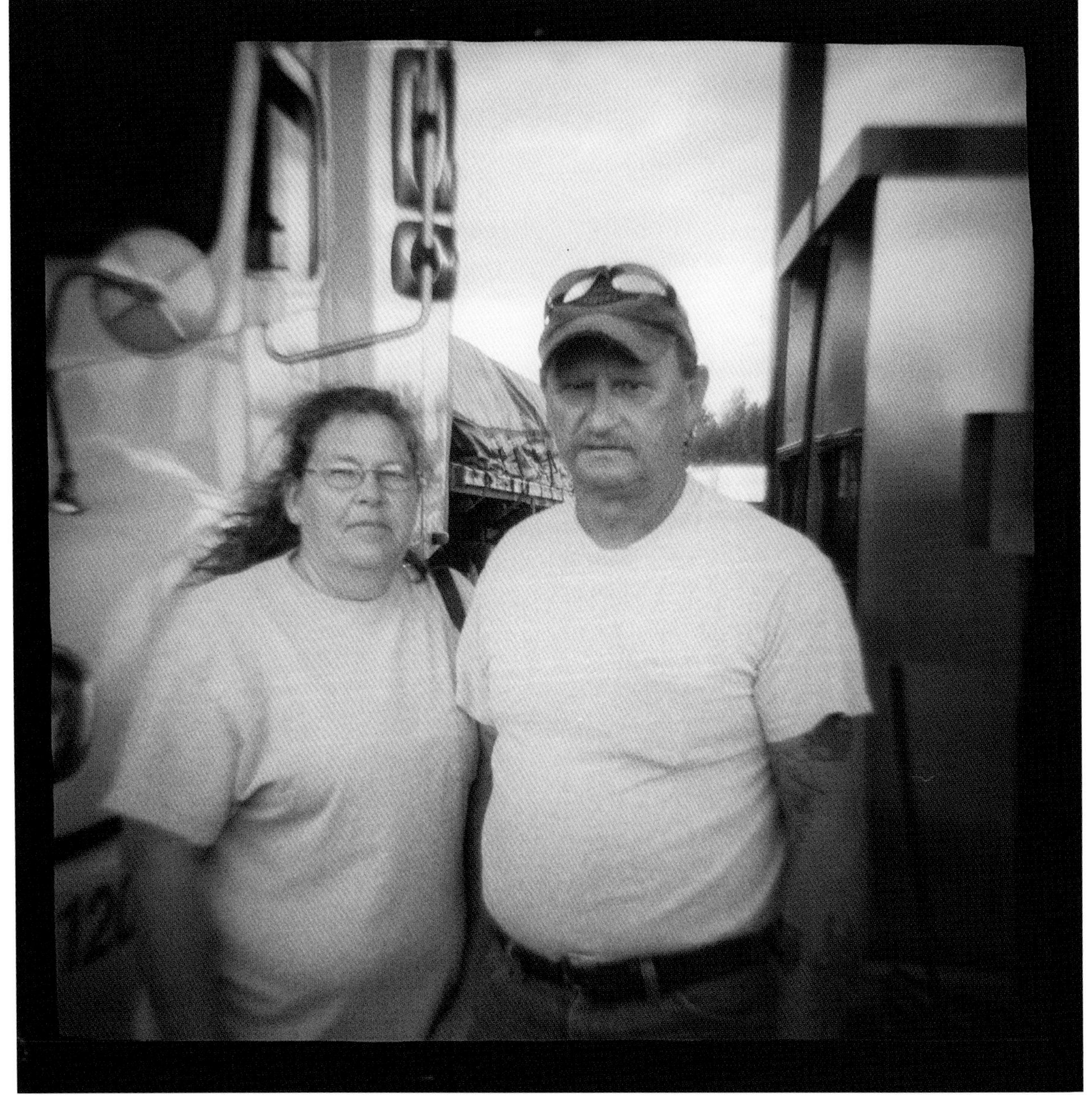

Husband and wife team; I-85, Exit 129

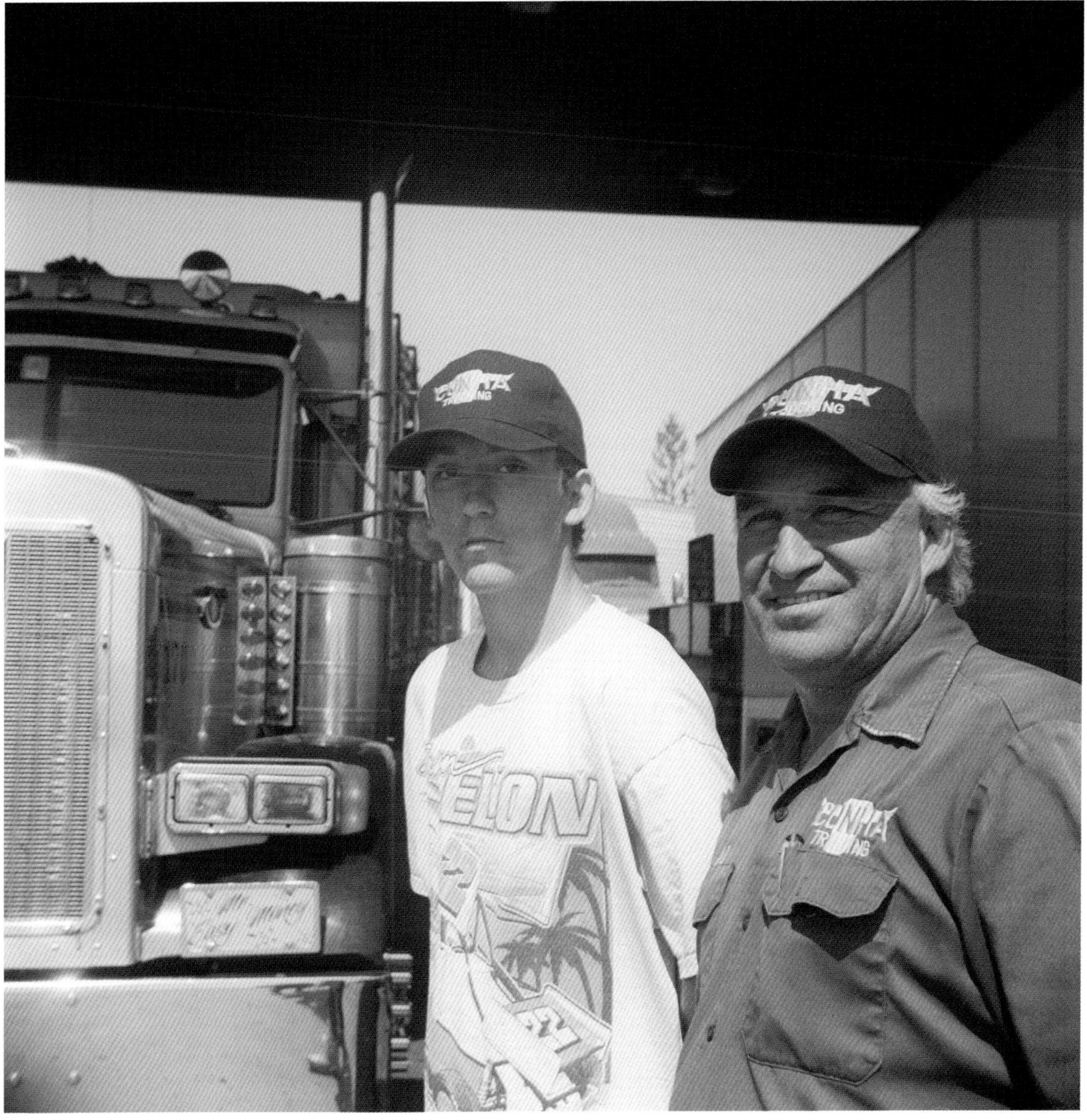

Father and son; I-80, Exit 176

41,000LBS. OF WATERMELON

Rhonda Sexton was a trucker from 1977 to 1981. "I quit for 23 years, had three kids, raised them and came back." She hangs dream catchers in the window and speaks with a slight lisp. Her tiny mouth vibrates round, subtly out of proportion to her square glasses. Today, she's taking 41,000 pounds of Florida watermelons to Detroit for 31¢ a mile. "I need 3,000 miles per week to make enough money. Last week I got 1,500. That's not enough to pay the bills." She says a 1,500-mile week pays about $250. The hours she might spend waiting to drop off a load are unpaid. At night, she counts her miles, adding up her earnings: the further away, the more she can provide. While driving, her thoughts stretch toward home, thinning in reach, thickening as a lump in her throat. "It gets lonely on the road. I miss my family."

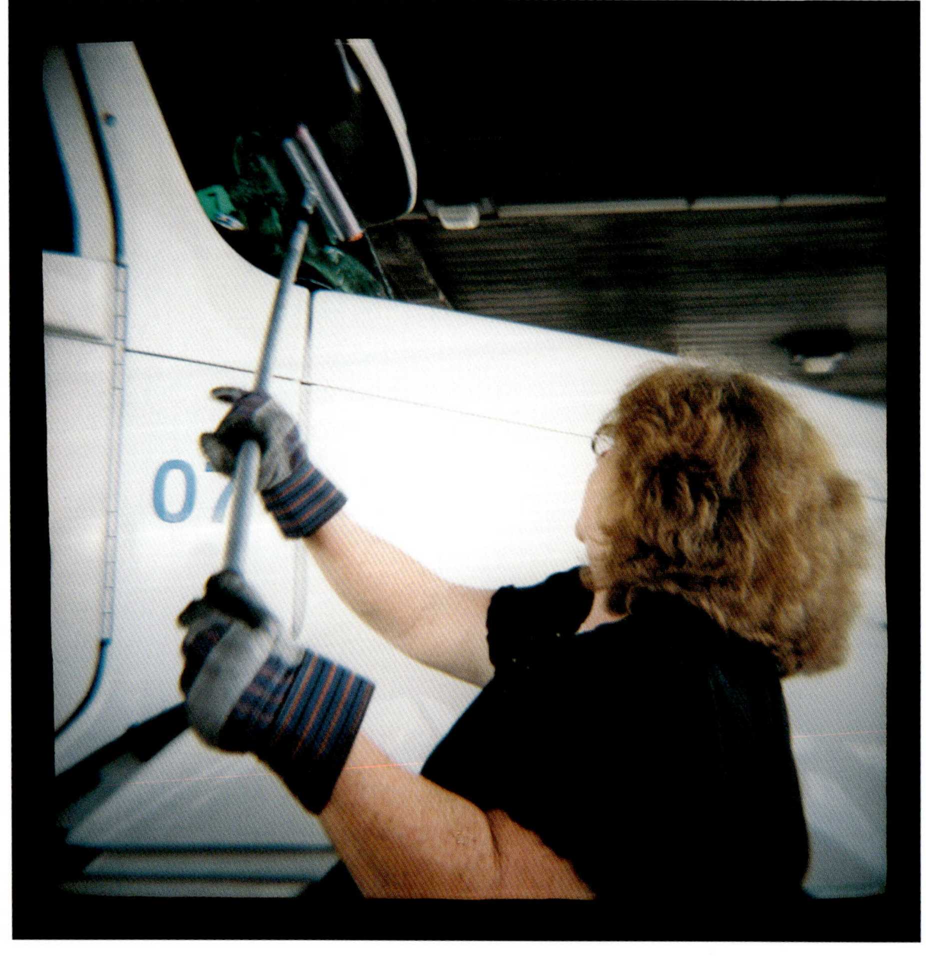

Rhonda Sexton; I-75, Exit 201

The Waiting

Condolous Reed refers to an uncommon silence. "You don't hear trucks running anymore fuel is so bad." Truckers idle to have power, air and heat while they are loading and unloading and while they sleep; it burns about a gallon of fuel every hour. His wife Mary rides with him and today she did not buy a five-dollar gallon of milk. He is waiting for an auxiliary power unit to be sent from the company; she is waiting to get home to their new baby, Elizabeth. She stands with her arms folded, nervously sweeping her hand across her left cheek. They have four other children at home. She tilts her head sideways, resting her right arm on top of her head. She repeats the same motion again and again, stopping to smile between.

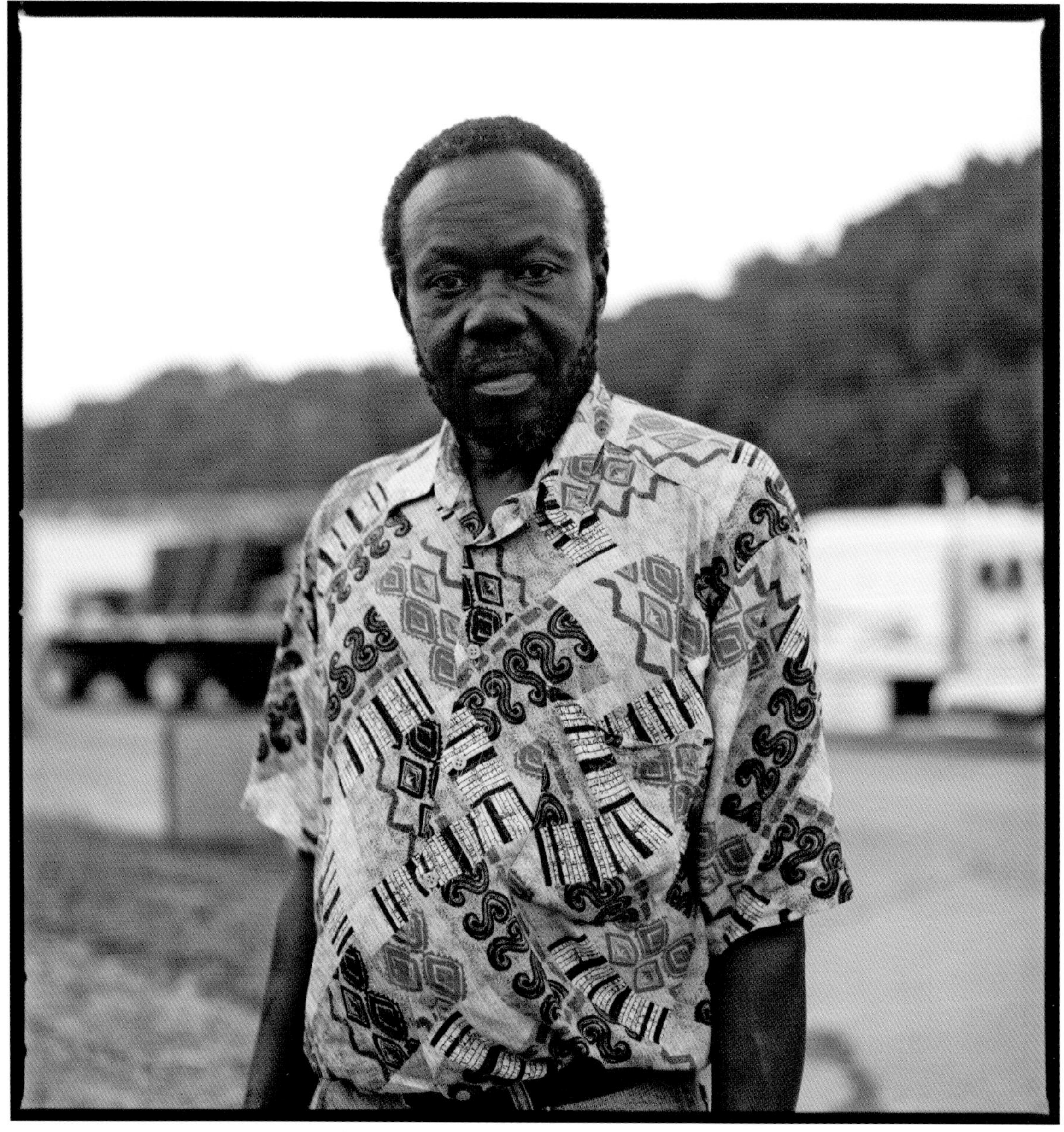

Trucker and Sudanese refugee; I-80, Exit 4

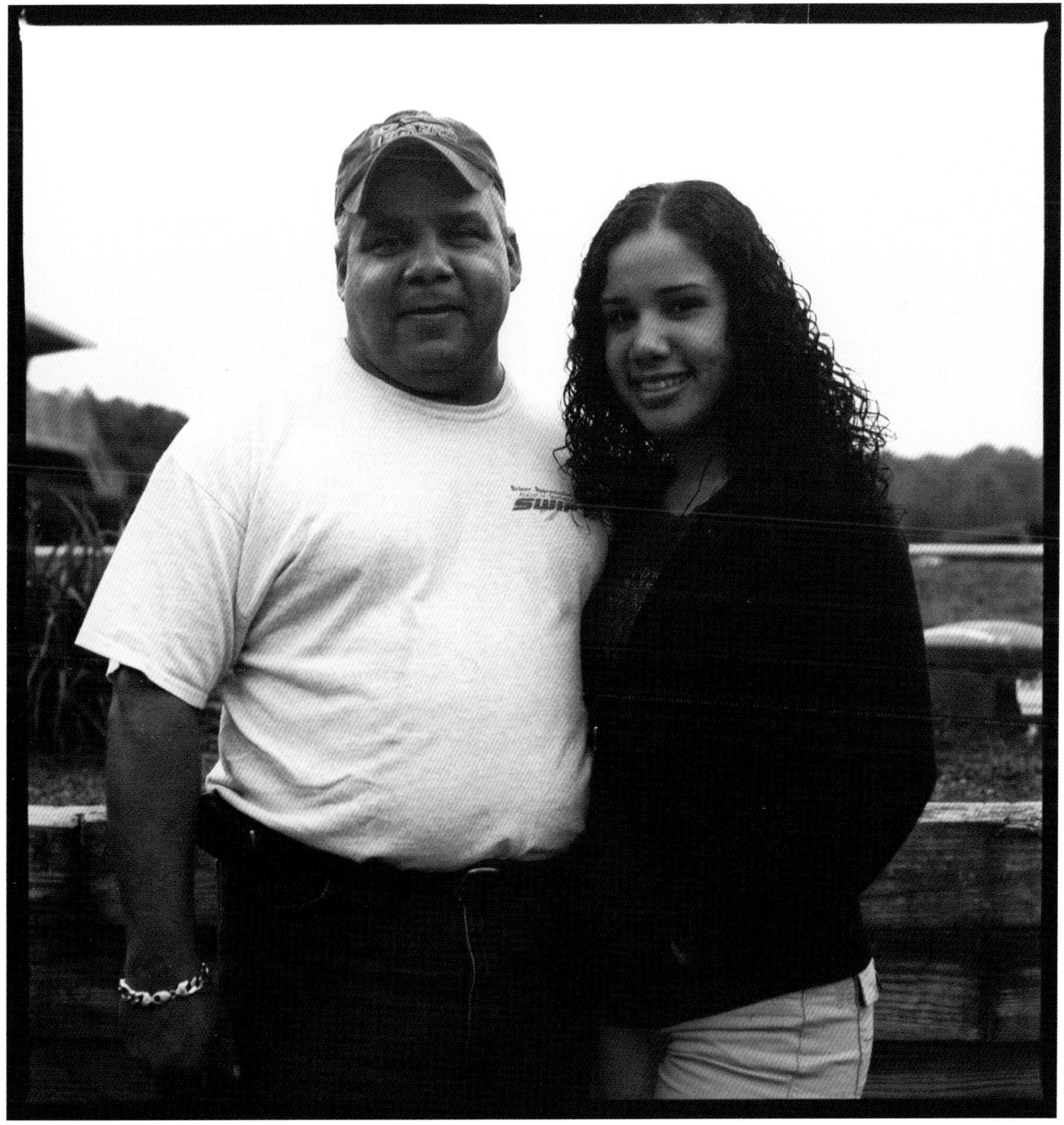

Omar Rivera and his daughter, who
is visiting him from Puerto Rico; I-80,
Exit 4

FUTURE

Amber, 20, will soon be married to Orlando, who is 21. They are both from Bakersfield, California. Orlando's father, all of his uncles and his godfather are truckers. So was Amber's mom. Orlando was hired seven months ago by one of the biggest trucking companies in the country. "They're so big, they lose track of you because you're just one driver out of so many. You try to get a hold of them and it takes a while." He communicates with them mostly through an onboard computer. "You type to them. They give us a load assignment. Every now and then when I really have a problem, like the other night I ran out of fuel, it took them four or five hours to get somebody out there."

Orlando's legacy cannot prepare him for all that can happen, and years of compounded loneliness cannot ease the burden of his absence on family at home, as Amber already knows, which is likely why she rides with him. "His mom tries to call us a lot. She gets real worried; she misses him like crazy." Their routes often take them near home, so they stop in to see her every few days, often just for a couple of hours. They gave up their apartment, wanting to save money for the future. Orlando seems to understand it is not something he can do forever. "It's good money, I guess...But you're gone all the time. If you have a family it wouldn't work." For now, it's just the two of them and a truck dog named Ladybug.

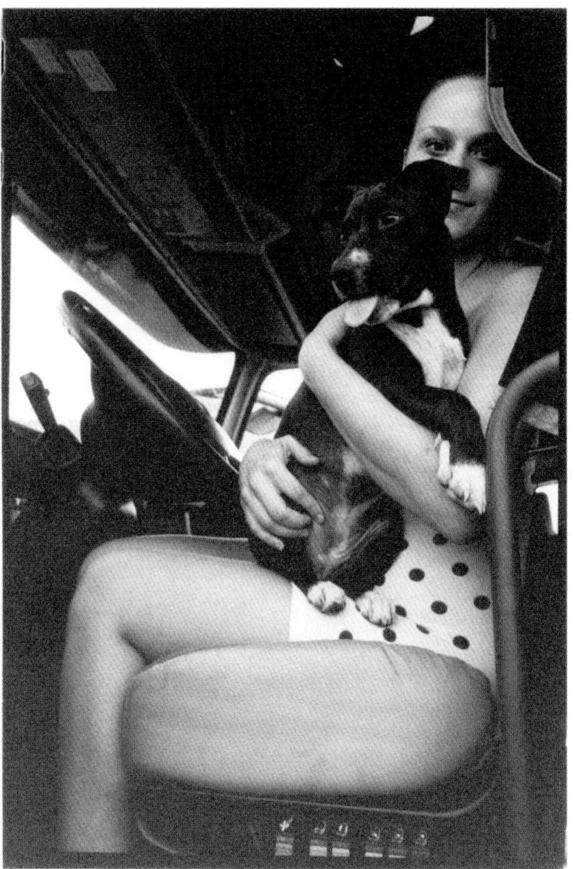

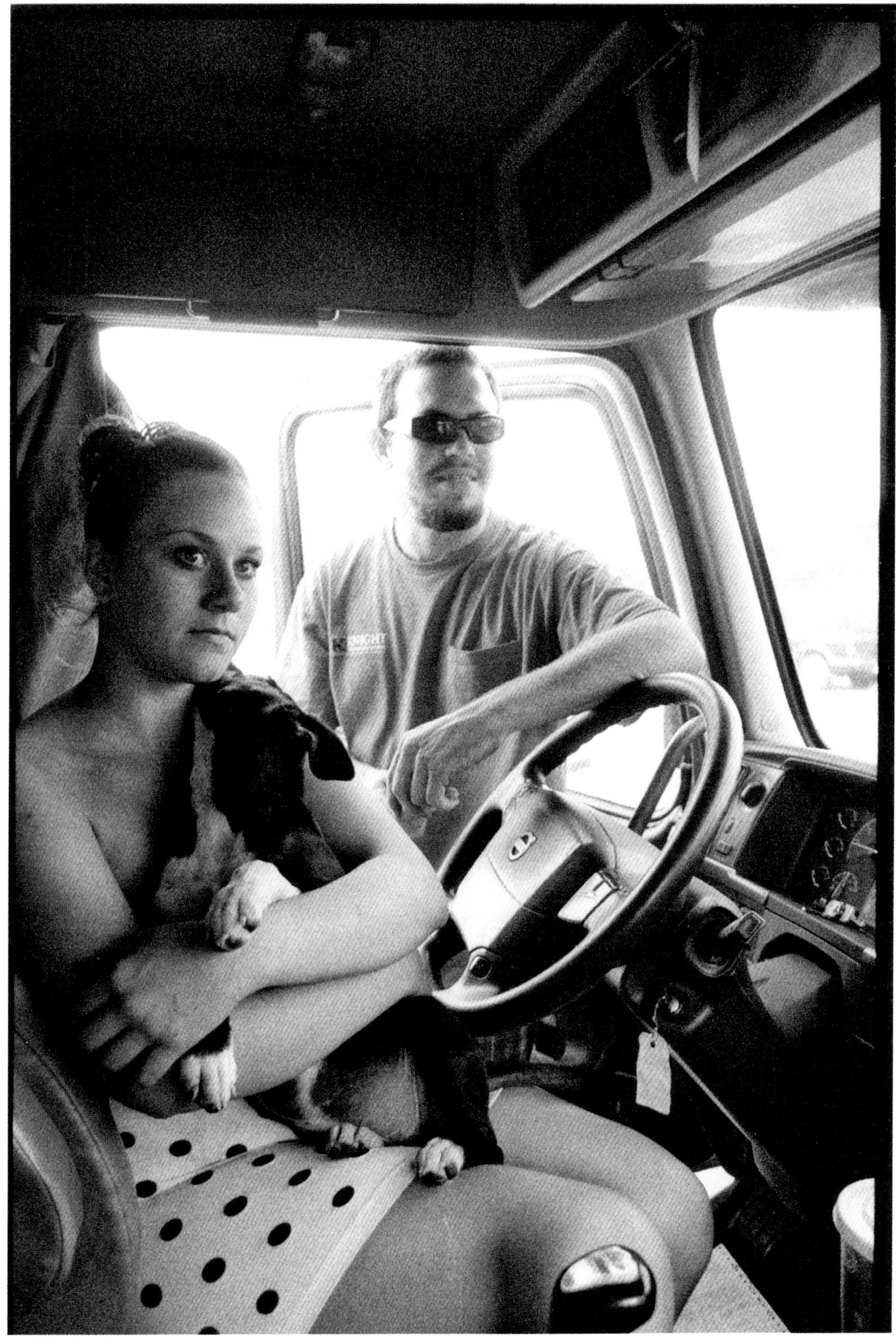

Amber, Ladybug and Orlando

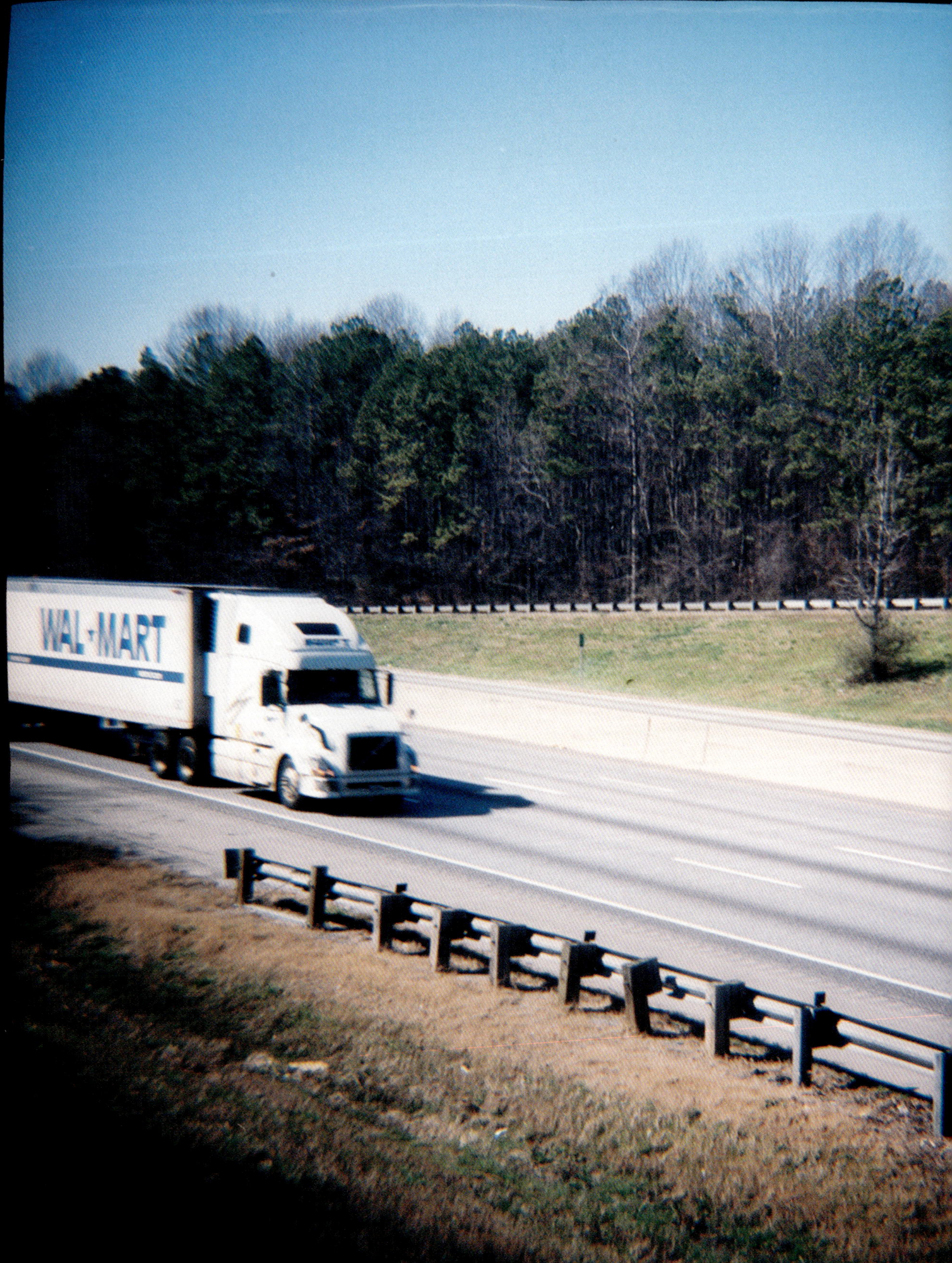
WAL-MART

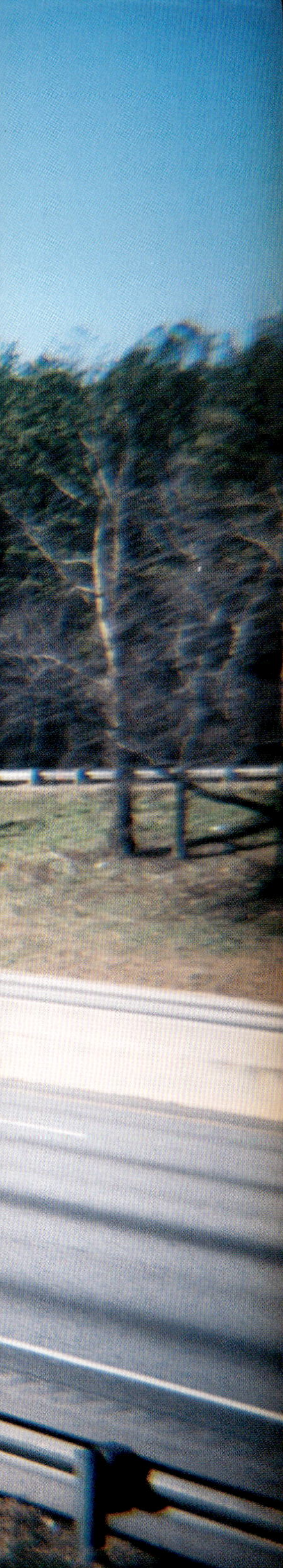

PHOTOGRAPHY

CREDITS

Photographers:

Phil Andrews was born in Greensburg, Pennsylvania, where he learned photography from his parents. Also a musician, songwriter and organizer, he is based in New York City.

Jenny Williamson is a photographer living in Decatur, Georgia. Her work is driven by her sentient nature and a desire to capture infinitesimal moments. She was inspired by this project and grateful for the experience.

Meshakai Wolf is a photographer and documentary filmmaker. He lives in Brooklyn, New York.

About the author:

Mary Richardson is a writer and editor. She lives in Atlanta, Georgia, where she publishes and edits a small arts quarterly, *FALSE* magazine. In 2006, she co-authored *New Orleans Bicycles*, also with Mark Batty Publisher. *Truckers* was inspired by the time she she spent as editor of a trucking industry trade magazine, an enlightening three years.

As required by the Federal Motor Carrier Safety Regulations Part 396.11

Date ________ Tractor No. ________________ ☐ I detect NO defect ☐ I detect the following defects:

INDICATE DEFECT AND SHOP, CITY & STATE WHERE REPAIRED.

____

____

OPERATOR'S SIGNATURE | MECHANIC'S SIGNATURE | SIGNATURE OF OPERATOR REVIEWING REPAIRS

TRAILER VEHICLE CONDITION REPORT

TRUCK NO. ________________

Date ________ Trailer No. ________________ ☐ I detect NO defect ☐ I detect the following defects:

INDICATE DEFECT AND SHOP, CITY & STATE WHERE REPAIRED.

____

____

OPERATOR'S SIGNATURE | MECHANIC'S SIGNATURE | SIGNATURE OF OPERATOR REVIEWING REPAIRS

TRACTOR VEHICLE CONDITION REPORT

As required by the Federal Motor Carrier Safety Regulations Part 396.11

Date ________ Tractor No. ________________ ☐ I detect NO defect ☐ I detect the following defects:

INDICATE DEFECT AND SHOP, CITY & STATE WHERE REPAIRED.

____

____

OPERATOR'S SIGNATURE | MECHANIC'S SIGNATURE | SIGNATURE OF OPERATOR REVIEWING REPAIRS

TRAILER VEHICLE CONDITION REPORT

TRUCK NO. ________________

Date ________ Trailer No. ________________ ☐ I detect NO defect ☐ I detect the following defects:

INDICATE DEFECT AND SHOP, CITY & STATE WHERE REPAIRED.

____

____

OPERATOR'S SIGNATURE | MECHANIC'S SIGNATURE | SIGNATURE OF OPERATOR REVIEWING REPAIRS

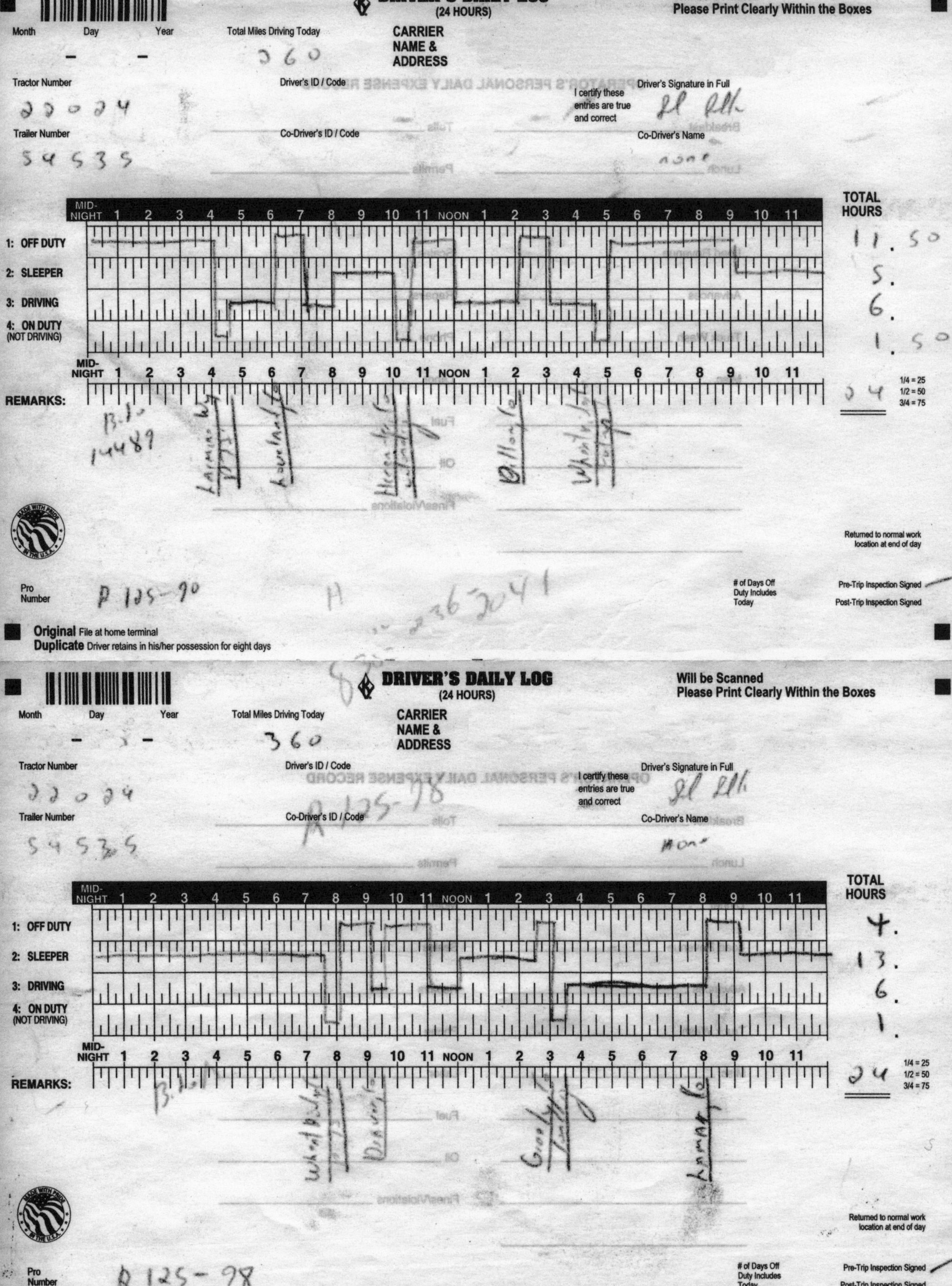

DRIVER'S DAILY LOG
(24 HOURS)

Will be Scanned
Please Print Clearly Within the Boxes

Month	Day	Year	Total Miles Driving Today	CARRIER NAME & ADDRESS
–		–	360	

Tractor Number: 22024
Driver's ID / Code:
Driver's Signature in Full:
I certify these entries are true and correct
Trailer Number: 54535
Co-Driver's ID / Code:
Co-Driver's Name: none

MID-NIGHT 1 2 3 4 5 6 7 8 9 10 11 NOON 1 2 3 4 5 6 7 8 9 10 11	TOTAL HOURS
1: OFF DUTY	11.50
2: SLEEPER	5.
3: DRIVING	6.
4: ON DUTY (NOT DRIVING)	1.50
	24

1/4 = 25
1/2 = 50
3/4 = 75

REMARKS:

B. Lo 14489

Returned to normal work location at end of day

Pro Number: R 125-90

# of Days Off Duty Includes Today

Pre-Trip Inspection Signed
Post-Trip Inspection Signed

Original File at home terminal
Duplicate Driver retains in his/her possession for eight days

DRIVER'S DAILY LOG
(24 HOURS)

Will be Scanned
Please Print Clearly Within the Boxes

Month	Day	Year	Total Miles Driving Today	CARRIER NAME & ADDRESS
–		–	360	

Tractor Number: 22024
Driver's ID / Code:
Driver's Signature in Full:
I certify these entries are true and correct
Trailer Number: 54535
Co-Driver's ID / Code: R 125-98
Co-Driver's Name: None

MID-NIGHT 1 2 3 4 5 6 7 8 9 10 11 NOON 1 2 3 4 5 6 7 8 9 10 11	TOTAL HOURS
1: OFF DUTY	4.
2: SLEEPER	13.
3: DRIVING	6.
4: ON DUTY (NOT DRIVING)	1.
	24

1/4 = 25
1/2 = 50
3/4 = 75

REMARKS:

Returned to normal work location at end of day

Pro Number: R 125-98

# of Days Off Duty Includes Today

Pre-Trip Inspection Signed
Post-Trip Inspection Signed